Systems for Planning and Control in Manufacturing

Systems for Planning and Control in Manufacturing

Systems and management for competitive manufacture

David K. Harrison

and

David J. Petty

Newnes

OXFORD AMSTERDAM BOSTON LONDON NEW YORK PARIS
SAN DIEGO SAN FRANCISCO SINGAPORE SYDNEY TOKYO

Newnes
An imprint of Elsevier Science
Linacre House, Jordan Hill, Oxford OX2 8DP
225 Wildwood Avenue, Woburn MA 01801-2041

First published 2002

British Library Cataloguing in Publication Data
A catalogue record for this book is available from the British Library

ISBN 0 7506 49771

For information on all Newnes publications
visit our website at www.newnespress.com

Composition by Genesis Typesetting, Laser Quay, Rochester, Kent
Printed and bound in Great Britain by MPG Books Ltd, Bodmin, Cornwall

Contents

Series Preface

'There is a time for all things: for shouting, for gentle speaking, for silence; for the washing of pots and the writing of books. Let now the pots go black, and set to work. It is hard to make a beginning, but it must be done' – Oliver Heaviside, *Electromagnetic Theory*, Vol 3 (1912), Ch 9, 'Waves from moving sources – Adagio. Andante. Allegro Moderato'.

Oliver Heaviside was one of the greatest engineers of all time, ranking alongside Faraday and Maxwell in his field. As can be seen from the above excerpt from a seminal work, he appreciated the need to communicate to a wider audience. He also offered the advice 'So be rigorous; that will cover a multitude of sins. And do not frown.' The series of books that this prefaces takes up Heaviside's challenge but in a world which is quite different to that being experienced just a century ago.

With the vast range of books already available covering many of the topics developed in this series, what is this series offering which is unique? I hope that the next few paragraphs help to answer that; certainly no one involved in this project would give up their time to bring these books to fruition if they had not thought that the series is both unique and valuable.

This motivation for this series of books was born out of the desire of the UK's Engineering Council to increase the number of incorporated engineers graduating from Higher Education establishments, and the Insitution of Incorporated Engineers' (IIE) aim to provide enhanced services to those delivering Incorporated Engineering courses and those studying on Incorporated Engineering Courses. However, what has emerged from the project should prove of great value to a very wide range of courses within the UK and internationally – from Foundation Degrees or Higher Nationals through to first year modules for traditional 'Chartered' degree courses. The reason why these books will appeal to such a wide audience is that they present the core subject areas for engineering studies in a lively, student-centred way, with key theory delivered in real world contexts, and a pedagogical structure that supports independent learning and classroom use.

Despite the apparent waxing of 'new' technologies and the waning of 'old' technologies, engineering is still fundamental to wealth creation. Sitting alongside these are the new business focused, information and communications dominated, technology organisations. Both facets have an equal importance in the health of a nation and the prospects of individuals. In preparing this series of books, we have tried to strike a balance between traditional engineering and developing technology.

The philosophy is to provide a series of complementary texts which can be tailored to the actual courses being run – allowing the flexibility for course designers to take into account 'local' issues, such as areas of particular staff expertise and interest, while being able to demonstrate the depth and breadth of course material referenced to a framework. The series is designed to cover material in the core texts which approximately corresponds to the first year of study with module texts focusing on individual topics to second and final year level. While the general structure of each of the texts is common, the styles are quite different, reflecting best practice in their areas. For example *Mechanical Engineering Systems* adopts a 'tell – show – do' approach, allowing students to work independently as well as in class, whereas *Business Skills for Engineers and Technologists* adopts a 'framework' approach, setting the context and boundaries and providing opportunities for discussion.

Another set of factors which we have taken into account in designing this series is the reduction in contact hours between staff and students, the evolving responsibilities of both parties and the way in which advances in technology are changing the way study can be, and is, undertaken. As a result, the lecturers' support material which accompanies these texts, is paramount to delivering maximum benefit to the student.

It is with these thoughts of Voltaire that I leave the reader to embark on the rigours of study:

'Work banishes those three great evils: boredom, vice and poverty.'

Alistair Duffy
Series Editor
De Montfort University, Leicester, UK

Further information on the IIE Textbook Series is available from
bhmarketing@repp.co.uk
www.bh.com/iie

Additional material, including slides with lecture notes, will be available at
www.bh.com/manuals/0750649771

Please send book proposals to:
rachel.hudson@repp.co.uk

Other titles currently available in the IIE Textbook Series

Mechanical Engineering Systems	0 7506 5213 6
Business Skills for Engineers and Technologists	0 7506 5210 1
Mathematics for Engineers and Technologists	0 7506 5544 5
Technology for Engineering Materials	0 7506 5643 3

Foreword

Modern industry requires high-quality engineers who can effectively deploy modern approaches to manufacture. Indeed a common complaint expressed by senior executives from industry is the difficulty of recruiting go-ahead people with appropriate skills. The Institution of Incorporated Engineers (IIE) therefore has a crucial role to play in providing the practical engineers and technicians needed by manufacturing enterprises.

Practitioners in modern manufacturing companies need more than technical knowledge. When I undertook my own apprenticeship in 1956, there was a strong emphasis on technical skills, to the exclusion of almost everything else. Even the theoretical courses, supporting the practical training, focused on how material would be physically processed. Today, engineers and technicians need a far wider range of skills. Not only are there a wider range of manufacturing technologies available, but people are typically called upon to undertake a far wider range of tasks. Traditionally, a young project engineer might be asked to redesign a jig or fixture. Today, he or she is more likely to be asked to devise a new system to support just-in-time manufacture. In addition, it is likely that it will be necessary for a technician/engineer to work alongside other professionals, perhaps from the Information Systems or Accountancy disciplines.

The manufacturing environment is changing and, in turn, practitioners need to adapt. There are five key areas in which industry is changing:

- **Product life cycles.** The rate at which new products are being introduced is increasing; particularly in consumer products such as cars or video equipment. This means it is unlikely that a production facility will manufacture the same product in the long term.
- **Focused factory.** There are several advantages in a small, focused manufacturing organization. This is leading many larger companies to adopt a decentralized approach, increasing the responsibilities of manufacturing professionals.

- **Global markets and competition.** With the lifting of trade barriers, competition is now fierce. There is enormous pressure on companies to improve key performance measures such as price, due date adherence and quality. At the same time, variety is increasing. This means innovative technologies and effective system design are becoming essential.
- **Reduced staffing levels.** In recent years, most companies have reduced employee numbers, particularly of indirect workers. Again, this means practitioners need a wider range of skills.
- **The factory as a system for manufacture.** A key theme in modern manufacturing companies is systems in the broadest sense. It is no longer pertinent to consider individual machines, but rather the manufacturing process in its entirety. Failure to take this view leads to the optimization of one part of the system at the expense of the whole system.

In summary, the modern engineer must have a broad skill base. This book therefore satisfies a crucial requirement by giving an overview of the different systems and management concepts necessary to work effectively in world-class manufacturing companies today.

Emeritus Professor Ray Leonard
BSc(Hons) PhD DSc CEng FIMechE FIEE FBIM
Formerly Director of Total Technology
University of Manchester Institute of Science and Technology

Introduction

This book is published under the auspices of the Institution of Incorporated Engineers. It will be useful for anyone with an interest in the application of management and systems (in the broadest sense) in manufacturing. Hopefully, this will prove a valuable text for engineering students at all levels. The material covered is also relevant to students undertaking business/management courses. Finally, it is hoped that the book is a useful reference source for practitioners.

The book is divided into four parts.

- **Introduction to the manufacturing enterprise.** This provides an overview of the manufacturing enterprise. It introduces the different ways companies can be classified and outlines the fundamental concepts of manufacturing layouts and basic principles of accountancy.
- **Systems and information technology.** This describes the main concepts of systems design. In particular, it covers structured systems analysis techniques and relational database management systems.
- **Quantitative methods.** This describes some of the common analytical tools that are employed to solve quantitative problems in manufacturing.
- **Management of manufacturing systems.** This describes the techniques that are used to plan and control a manufacturing organization.

In addition to these, there are also sections containing exercises to support the main text.

Over recent years there have been a number of changes in the world economy, for example large manufacturing corporations are now highly decentralized and operate globally. In general, the basis upon which businesses compete has changed and one of the effects of this is that information systems now play a central coordinating role in the operation of such businesses (this is discussed in detail in Part 2). It is for this reason that information technology/information systems (IT/IS) is featured prominently in this book.

D. K. H. and D. J. P.

Part 1

The Manufacturing Enterprise

Automation means efficiency?

January

The board of GenerEx sat listening to the presentation of Robert Gould, the company's manufacturing director. GenerEx was a well-established manufacturer of small electric generators based in England. The company had had a difficult year – profits were low and for the first time in ten years, there had been redundancies. There was increasing competition from abroad particularly from Eastern Europe. Robert was pleased today, however; he'd been working hard on developing a proposal for a new automated winding line. The generator core was a key component for the company and represented most of the cost of the finished product. The current method of manufacture was very labour intensive. The company employed hundreds of shop floor workers who wound copper coils using simple hand-turned formers and then inserted these into a pack made of sheet steel. The work was quite skilled. The company made a variety of different sized cores to provide a range of power outputs. Some of the cores were standard and made in relatively large quantities. Others were made specifically to meet the needs of customers and, as a result, were made in relatively small volumes. The automated core manufacturing line that Robert was proposing would revolutionize production.

Robert drew his presentation to a close. 'So to conclude, if the board agrees to fund my proposal we will be able to reduce labour costs enormously. The new line would be able to produce cores at the rate of 12 a minute. It is fully automated and the only labour required for its support would be a maintenance engineer and three unskilled operatives to load materials and unload finished cores. Automation means efficiency. At full production, we estimate that the line will be able to manufacture over one million cores per year which represents almost 40% of total requirements.'

The board unanimously supported the capital expenditure for the new line. The next item of business was a statement by the marketing director, Alan Ashman. He outlined the problems the company was experiencing with overseas competition. 'There is no way we can compete on price for standard generators. The new automated line will help, but as Robert has said, it will only make 40% of our requirements. Labour costs in Eastern Europe are far lower than here in the UK and therefore I propose that we focus on the non-standard generators. With our engineering skills, we can offer a service to clients who need a specialized generator. We have always had higher profit margins for non-standard generators. With this new strategy, I anticipate our production volumes to fall. In fact, the new automated line should be able to produce 65% of our total requirements. Overall, however, we should be able to increase profits considerably.'

Again, the board unanimously supported the proposal to focus on specialized generators.

June

Robert Gould inspected the newly installed core production line with his chief production engineer, Steve Curbishley. The line had been purchased from a German company and it was well engineered. Commissioning had gone according to plan and production was due to start on Monday. Robert said, 'This is an impressive piece of kit. Considering the complexity, it's amazingly reliable.' 'I agree', said Steve, 'it's well made. Pity it takes a couple of days to change from one core to another. I think we can get this down to a day in the long term, when we're used to dealing with such complex equipment. The previous technology was pretty simple because it depended on operator skill.' The problem they did not discuss was the early retirements that were necessary. Even though the company had not actually made anyone redundant, many of the operators were not happy and felt they had been forced to accept early retirement. Despite the fact that an enhanced pension and a lump sum had been offered, morale was low. 'Well the sooner we can get this thing on-line the better', said Robert. 'We need to start making some money considering how much we've spent.'

December

The managing director, Keith Johnson, was furious and had called a crisis meeting of the board. 'What the hell is going on. I've spent over £1 M on an automated line plus £250 K on labour restructuring. It also cost almost £150 K to implement the new strategy on specialized motors. Somehow, we're in a bigger mess than before! I have to tell you, gentlemen, that the finance director and I had a very difficult meeting this morning with the bank. Our cash flow is absolutely terrible and the bank is close to pulling the plug. Could someone explain to me why we're on the point of insolvency?' James Dickinson, the finance director, looked at the papers in front of him: 'stocks of generator cores are through the roof – almost literally. I would have thought that the new line would have reduced stocks bearing in mind it's the latest technology. What's the problem, Robert? Is there a problem with the line?' The manufacturing director sat nervously in his seat. The truth was that he didn't really understand what had gone wrong. Everything had happened so quickly. 'The line's working fine. The problem is that we have to make large volumes of product because it takes between one and two days to change the line. Marketing are bringing in bits and bobs of specialized orders. Why can't they bring in some decent volume.' 'As I recall, Robert', said Alan Ashman, 'you voted in favour of the new marketing strategy. There's another problem, we haven't supplied some of our customers, particularly with the smaller volume orders and they're furious. We'll have to rehire some the people who took early retirement as a temporary measure. I afraid that's going to cost us. The fact of the matter is that the new line has been a disaster from start to finish.' All eyes turned to the manufacturing director who was looking extremely uncomfortable. He looked around the table. 'As I recall, we all voted in favour of investing in the new line!'

Analysis

It might be thought that the only thing that influences the type of manufacturing technology a company uses is the nature of the product. While the product must be taken into account, the relationship of the company to its market is equally important. In the case of GenerEx, the manufacturing technology chosen was effective at making the products required. The problem was the new line, with its long set-up time, was really suited to volume production. The mismatch of manufacturing and marketing strategies brought the business to the brink of collapse. This will be discussed further under 1.6 in Chapter 1.

Summary

Every engineer needs to understand manufacturing at the macro level. This part places the manufacturing function into context by considering the organization as a whole. Three areas are discussed: first, how manufacturing systems are designed and how they support business strategy. Second, financial and management accounting are discussed (all engineers need to have a good understanding of finance if they are to obtain a senior management position). Finally, the management of the design function is discussed and how it relates to manufacturing.

Objectives

At the end of this part, the reader should:

- Appreciate the structure of manufacturing organizations at the top level.
- Understand how manufacturing systems are planned and designed strategically.
- Understand the financial concepts that underpin the operation of any organization, including manufacturing organizations.
- Understand how the design function is critical within manufacturing organizations.

1 Manufacturing systems

1.1 Overview

Manufacturing is fundamentally concerned with transforming raw materials into finished products for sale. This activity cannot be divorced from the economic reality of wealth creation. Indeed, most governments in the developed/developing world are acutely aware that prosperity is closely linked to the ability to export goods. The prosperity of the UK is crucially dependent on the ability of manufacturing industry to be competitive. This is for two reasons: first, to win export orders and thus generate revenue. Second, to sell products domestically at the expense of imports in order to achieve a favourable balance of trade. In the past, this second objective could be achieved by government intervention with the imposition of import duties. With liberalization in world markets, companies have no refuge from competition.

Prior to the industrial revolution, manufacturing was not undertaken systematically. Artefacts were made by craftsmen working either alone or in small groups. Factories in the modern sense did not exist. After the industrial revolution, the nature of manufacturing changed. By 1850, most production took place in factories. While manufacturing technology has improved enormously, the factories of the mid-nineteenth century would be immediately recognizable to modern engineers. Factories normally consist of a number of individual *processes* linked into a coherent *manufacturing system* (see Figure 1.1).

The manufacturing system itself is part of a wider organization. Traditionally, the manufacturing system has been supported by a number of functional departments. The activities of these departments (and manufacturing itself) are coordinated by *production planning* over the short to medium term. Note also that there are links between the organization and the outside world. The entire system for transforming basic raw materials into products for end users is referred to as the

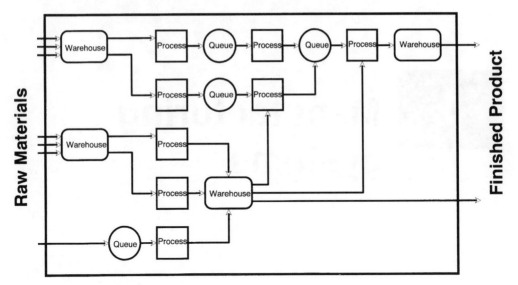

Figure 1.1   *A typical manufacturing system*

supply chain (see Figure 1.2). The functional departments are as follows:

Sales	Responsible for defining demand.
Design/Development	Responsible for specifying the item to be manufactured.
Production Engineering	Responsible for defining how the item shall be made.

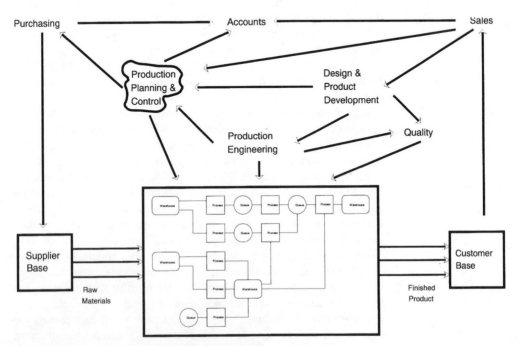

Figure 1.2   *A typical manufacturing organization*

Quality   Responsible for ensuring product conforms
                        to requirements.
Production Planning   Responsible for defining when product shall
                        be made and in what quantity.
Purchasing   Responsible for generating/expediting
                        purchase orders and choosing suppliers.
Accounts   Responsible for managing the company
                        finances.

The objectives of a manufacturing management can be summarized as supporting the production of items of appropriate quality in the appropriate volumes at the right time at optimal cost. It is concerned with such issues as planning, plant design and the broader supply chain. It is *not* directly concerned, however, with issues such as design, R&D, marketing or the process technology itself.

In recent years, manufacturing management techniques have been the focus for much attention in industry. Various philosophies such as manufacturing resource planning (MRPII), just-in-time (JIT) and optimized production technology (OPT) have been fashionable at different times. None of these, however, are panaceas for success. It should be recognized that certain issues within an organization *cannot* be addressed simply be improving methods of manufacturing management. For example, it is a commonly quoted maxim that 80% of product cost is fixed at the point of design. Organizations with poorly designed products cannot be transformed into profitable enterprises simply by the introduction of *any* manufacturing management technique. Similarly, it is often the case that some problems can only be addressed by improvements in the manufacturing processes.

1.2 Hierarchy of management control

Planning and control is concerned with the ongoing management of the manufacturing organization. This can be considered to be a hierarchical process as shown in Figure 1.3.

1 **Corporate planning.** This is concerned with the overall strategy and direction for the business as defined by senior management. The timescale for this activity is in the order of years.
2 **Aggregate planning.** This is tactical planning over a timescale of months to years. It is concerned with ensuring that the organization is capable of meeting corporate objectives.
3 **Master planning.** This defines a time phased plan for manufacture, normally in terms of high-level assemblies. The timescale for this activity is typically in the order of weeks.
4 **Intermediate planning.** This is a plan of items to be manufactured or procured. In most cases, this is derived from the master plan. Timescales are in weeks or possibly days.

Figure 1.3
Hierarchy of control systems

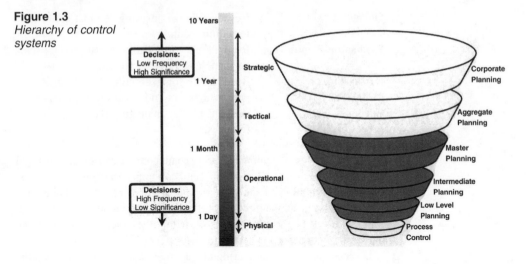

5 **Low-level planning.** This is a detailed plan defining what activities will actually take place at a moment in time. Timescales are measured in days or less.

6 **Process control.** This is concerned with the actual control of processes in real time.

Most of this module will address the operational (shaded in Figure 1.3) levels of manufacturing management.

1.3 Classification of manufacturing organizations

Defining an appropriate strategy for manufacturing management will depend on the nature of the organization. For this reason, many authors have attempted to classify manufacturing organizations. There are several ways in which an organization can be classified:

- **Size.** Large, medium or small enterprise.
- **Type of product.** Chemical, metal working, electronic assembly, etc.
- **Technology.** Manual, mechanized, etc.
- **Machine layout.** Process oriented (functional), product oriented, cellular.
- **Market.** Make to stock, make to order, etc.
- **Material flow.** Job, batch, mass, flow.

The manner in which companies serve the market has a particularly important influence on the design of manufacturing management systems. The following types exist:

Make to stock (MTS). Customers require products with almost no forward notice. This means that the supplier manufactures in *anticipation* of demand, maintaining stocks of finished product.

Configure to order (CTO). Customers are prepared to wait for a limited period, but not long enough to allow manufacture from base raw materials. The company manufactures items in anticipation of demand, but only *configures* these subassemblies or components into saleable products when actual customer orders are received. This mode of manufacture is extremely common in industry.

Make to order (MTO). In this case the company does not commit resources until firm customer orders have been received. In the purest form, this mode of manufacture is extremely rare.

Engineer to order (ETO). This is similar to MTO except that in this case, the item to be manufactured is defined by the customer. Thus, the supplier needs to interpret the requirements of the customer before manufacture can start.

Jobbing manufacture is characterized by one-off production. Jobbing manufacturers have unspecialized, flexible plant. Levels of automation are low, with highly skilled operators. Profit margins are high. Mass production is highly automated, specialized production at high through-put with low variety. Profit margins are low, so utilization of plant is very important, particularly as mass production usually involves large investment. Batch manufacture has intermediate characteristics except in one area – manufacturing management complexity. In batch manufacture, production control can be highly complex.

There are a variety of techniques that can be applied to different types of companies. Critical path analysis (CPA), for example, is widely applied in complex jobbing production (e.g. shipbuilding). This chapter will focus of the type of system that is prevalent in the batch manufacture of discrete items. Nonetheless, some of the concepts covered are common to all manufacturing organizations.

1.4 Strategy

All commercial organizations need an overall strategy. A number of terms are used, often interchangeably, to describe activities associated with this activity; corporate planning, strategic planning, strategic management, corporate strategy and so forth. All of these, however, have a common theme. They aim to define an overall direction for the business at the highest level. Strategy in itself is hard to define. By looking at specific cases, however, it is clear that different organizations have different strategies. Mercedes-Benz and Daewoo both manufacture cars, but both clearly have entirely different strategies. This section will discuss strategy and, in particular, how this relates to the manufacturing function. For the purposes of this book, the following definitions are adopted:

Corporate strategy. This is an all-encompassing and all-embracing direction for the complete entity (e.g. a corporation). This provides top-level guidelines that direct all other operations within the entity. An example of corporate strategy is determining by which parts of the world it is prepared to do business.

Business strategy. Guided by the corporate strategy, this seeks to maximize benefits for the stakeholders. Typically, in practice, this is financially driven. For example, some manufacturers of computer printers sell their machines at cost in the expectation of generating revenue from sales of replacement ink cartridges.

Manufacturing strategy. This is concerned with producing products in the most effective manner possible.

1.5 Developing a strategy

The strategic function for commercial organizations is illustrated in Figure 1.4. The steps in the process are as follows:

1 This defines the overall business direction. Often this will be embodied in a mission statement.
2 It is necessary to assess the organizational status; i.e. how well is the business performing overall?
3 A statement of objectives needs to be drawn up. This will be far more specific than a mission statement and includes targets.
4 Detailed plans will need to be established to determine how the objectives will be met. These will be broken into specific strategies for the different functions of the business.
5 & 6 The plans then need to be actioned and performance monitored.

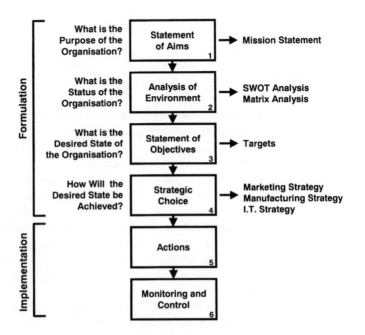

Figure 1.4 *The strategic process*

1.6 Manufacturing strategy

It is clear that manufacturing strategy is crucial to competition. If the basic design of the manufacturing system is poor, no matter how well it is managed, the company will not be able to compete. It is essential therefore that manufacturing strategy is derived from business objectives. Hill (2000) proposed the well-known model for defining manufacturing strategy as shown in Figure 1.5.

Figure 1.5
Manufacturing systems strategy

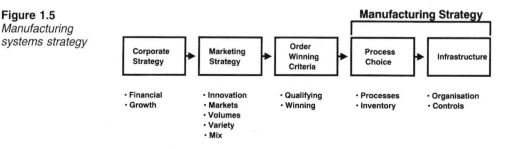

This model starts from the definition of a corporate strategy. In particular, targets need to be defined in financial and growth terms (e.g. projected sales and profit levels). This feeds into the market strategy that defines the way in which products will be sold. This addresses issues such as whether new products need to be designed (innovation) and if new markets need to be developed. From a manufacturing point of view, the volume, variety and mix of the product range are particularly important as this drives the process choice.

1.7 Process choice

Within the general area of metal working manufacture, there are a number of different types of production:

Individual machine. Manual loading and unloading. Job or small batch manufacture with functional layout. Example – subcontract manufacture.

Flexible manufacturing cell (FMC). Individual CNC machines linked together, often by a robot. Example – aerospace manufacture.

Flexible manufacturing system (FMS). CNC machines linked in a flow arrangement. Often uses a conveyor with some form of standard pallet or work holding device. Example – machine tool manufacture.

Batch flow line (group technology). Linked machines with parts grouped together in batches to minimize the effect of set-up times. Example – brake lining production.

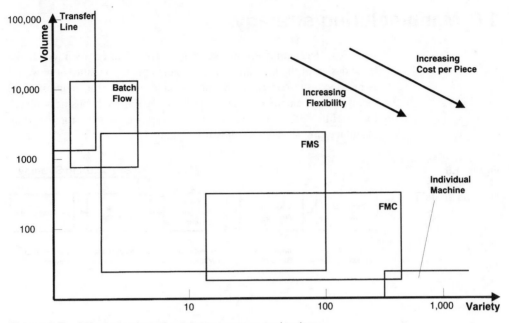

Figure 1.6 *Effect of volume/variety on process technology*

Transfer line. Highly automated, high volume production with single piece flow. Example – car production.

It is critical that the process choice matches the overall strategy for the organization (see the GenerEx case study at the start of this part and also the graphical representation shown in Figure 1.6).

1.8 Facilities organization

The problem of facilities organization and layout is common to almost every enterprise. Shops, restaurants and offices need to carefully consider how facilities will be arranged. In a manufacturing context, there are four types of layout:

Fixed material. In this case, the plant moves around fixed products under manufacture. This is typically employed for large products (e.g. shipbuilding or civil engineering projects).

Functional. Similar machines are grouped together (e.g. subcontract job-shop).

Group technology. Machines are arranged to manufacture product families (e.g. brake lining manufacture).

Flow/Product. Here, the production facilities are designed to manufacture specific products in large quantities (e.g. volume car product).

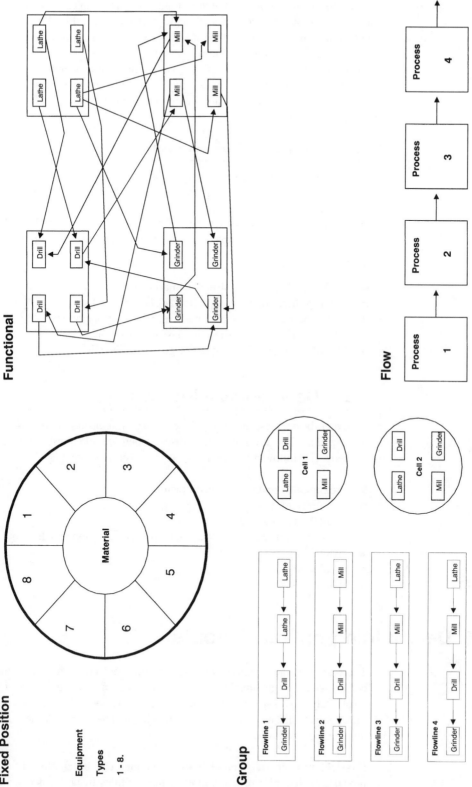

Figure 1.7 *Types of layout*

These forms of layout are shown in Figure 1.7. The choice of which of these approaches to adopt is a strategic decision. Once this choice has been made, it is the task of the manufacturing engineer to lay out the manufacturing facilities in the most appropriate manner. It should be noted that most of the work of the manufacturing engineer is directed towards the modification of an existing factory, rather than starting from a greenfield site or undertaking a complete reorganization.

1.8.1 Hierarchy of facilities

Most factories are organized hierarchically with a corresponding management structure. This approach is very common and is applied in many organizations. This philosophy allows management to understand and control a complex environment. The nature of the hierarchy will vary from company to company but in general, the larger the company, the greater the number of levels in the structure.

- **Divisions**. Services a particular market.
- **Factory**. Manufactures a particular class of product.
- **Department**. Undertakes a particular class of operations.
- **Group**. Undertakes a specific class of operations.
- **Machine**. Undertakes a specific group of operations.

1.8.2 Geographical location

In larger businesses, it may be necessary to consider a multi-site organization. The question then arises, where should these sites be located? The following factors need to be taken into account:

- Proximity to materials/subcontractors
- Proximity to markets
- Availability of space
- Quality of infrastructure (transport, power, telecommunications)
- Availability of labour
- Availability of grants
- Political stability.

1.9 Determining optimal layouts

There are several objectives that need to be considered when carrying out a layout design: maximizing flexibility, minimizing space, etc. In most analyses, however, the primary objective is normally to minimize the aggregate distance travelled by components. There are four methods that can be employed:

Physical modelling. A number of techniques can be employed: 3D models, cardboard representations of the facilities to be located. One popular technique is to use scale drawings with pins to represent the location of facilities. String can then be used to model the information flows.

Computer heuristic methods. There are a number of computer programs that produce an optimal arrangement of facilities. The best-known package is Computerized Relative Allocation of Facilities Technique (CRAFT). There are, however, a variety of alternatives.

Simulation methods. These allow the engineer to observe the consequences of a particular layout choice under differing conditions. The great virtue of simulation is that it permits the analysis of problems too complex to be handled by other means. Modern simulation packages are highly sophisticated and allow realistic models of manufacturing systems to be produced. With powerful computers, it is possible to generate animated 3D views of a manufacturing system using virtual reality techniques. This will be discussed further in Chapter 14.

Cross and relationship charts. These are manual techniques for defining an appropriate layout. The computer heuristic techniques discussed above, however, utilize some of the ideas embodied in cross charts. These are discussed in detail below.

1.9.1 Cross and relationship charts

A cross chart defines the material flow patterns within a manufacturing system in tabular form. An example of a cross chart is shown in Figure 1.8.

The numbers in the cross chart represent the number of batches of products moving from one resource to another in a given time period.

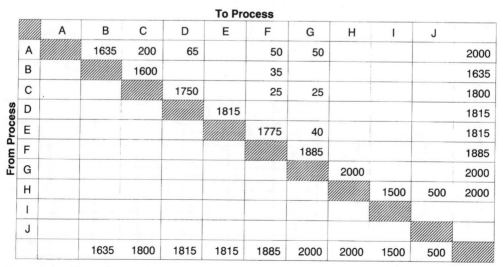

		A	B	C	D	E	F	G	H	I	J	
	A		1635	200	65		50	50				2000
	B			1600			35					1635
	C				1750		25	25				1800
	D					1815						1815
	E						1775	40				1815
	F							1885				1885
	G								2000			2000
	H									1500	500	2000
	I											
	J											
			1635	1800	1815	1815	1885	2000	2000	1500	500	

Figure 1.8   *Example of a cross chart*

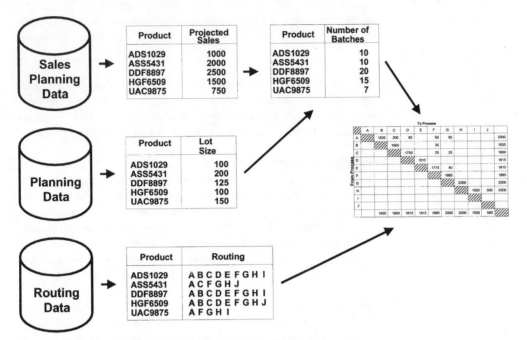

Sales Planning Data

Product	Projected Sales
ADS1029	1000
ASS5431	2000
DDF8897	2500
HGF6509	1500
UAC9875	750

Product	Number of Batches
ADS1029	10
ASS5431	10
DDF8897	20
HGF6509	15
UAC9875	7

Planning Data

Product	Lot Size
ADS1029	100
ASS5431	200
DDF8897	125
HGF6509	100
UAC9875	150

Routing Data

Product	Routing
ADS1029	A B C D E F G H I
ASS5431	A C F G H J
DDF8897	A B C D E F G H I
HGF6509	A B C D E F G H J
UAC9875	A F G H I

Figure 1.9   *Generation of a cross chart from a database*

Thus, in Figure 1.8, 1635 batches move from resource A to B (note that no batches move from B to A). Cross charts are derived from projected sales, planning data and routings. In most modern organizations, it will be possible to produce a cross chart from the company database. The mechanics of this process are illustrated in Figure 1.9.

The chart is generated by taking a sales forecast and converting this into the number of batches by using the lot size. Process routings can then be used to calculate the number of movements between processes.

Cross charts are useful for visualizing material flow through an organization. For example, the cross chart in Figure 1.8 shows that the flow through the manufacturing system is undirection. Cross charts can be used quantitatively to minimize the number of non-adjacent movements. Consider the simple plant layout and cross chart in Figure 1.10.

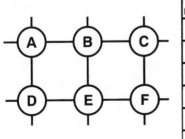

From \ To	A	B	C	D	E	F
A		150	75			
B			300	75		
C	90	75		60		
D		150			200	150
E		75				
F		150		150		

Figure 1.10   *Simple plant layout and associated cross chart*

Figure 1.11
Revised layout

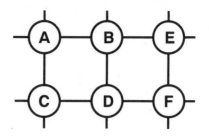

In Figure 1.10 for example, there are non-adjacent movements for A to C of 165 (75+90) and D to F of 300. By reorganizing the plant, these non-adjacent loads can be minimized as shown in Figure 1.11.

In practice, it is necessary to calculate all of the non-adjacent loads as shown below:

$$
\begin{array}{lll}
\text{A–C} & 75 + 90 & = 165 \\
\text{D–F} & 150 + 150 & = 300 \\
\text{A–F} & 0 & \\
\text{D–C} & & = 60 \\
\end{array}
$$

In complex situations, many iterations will be needed to establish an optimal solution and a computer will probably be required.

1.10 New facilities

A more typical task for the manufacturing engineer is to introduce another entity into an existing environment. Variations of the above can be used, but it is also possible to consider this class of problem analytically. Consider Figure 1.12: this represents a factory with four existing facilities. The question arises, at what (x,y) position P should the new facility be located such that the aggregate distance between workcentres is minimized?

The distance between any point P and an existing facility F_i is given by the following:

$$
\text{Dist} = \sqrt{(x - x_i)^2 + (y - y_i)^2}
$$

The aggregate distance can be calculated at any point using the equation below.

$$
\text{Tot dist} = \sum_{i=n}^{i=1} \sqrt{(x - x_i)^2 + (y - y_i)^2} \tag{1.1}
$$

This can be plotted over the available area (as shown in Figure 1.12). The distances between the workcentres can be weighted to allow for the relative importance of proximity between the facilities. In practice, there

Figure 1.12
Layout example

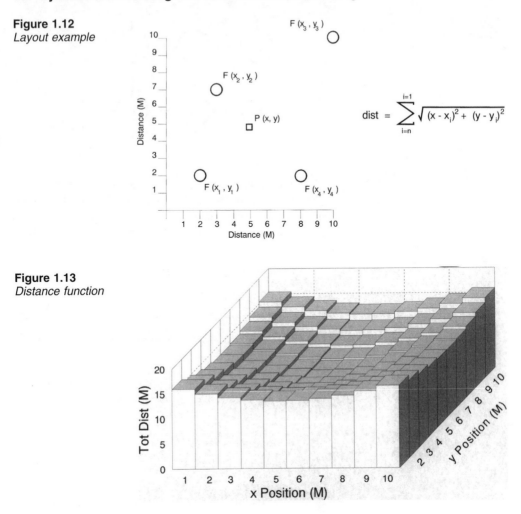

$$\text{dist} = \sum_{i=n}^{i=1} \sqrt{(x - x_i)^2 + (y - y_i)^2}$$

Figure 1.13
Distance function

will be only a limited number of positions the facility can be located. Simple spreadsheets can be used to automate the calculations.

1.11 Group technology (GT)

1.11.1 Functional layout

The traditional method of layout in batch manufacturing is functional grouping. In this approach, machines of similar types are grouped together. Material flow between the individual machines is defined by process routings and is highly complex. This form of layout has a number of limitations:

● Considerable distance travelled by components.
● High levels of material handling.
● Queuing is uncontrolled.

- High work in process (WIP).
- Long lead time.
- Difficult to understand the system.
- Difficult to simplify.

1.11.2 Development of GT

Many of the ideas behind GT date back to the 1920s. The concepts were refined and applied in the UK in companies such as Serck-Audco, Ferranti and Ferodo during the 1960s. Much of the pioneering work was undertaken by institutions based in the Manchester area: Salford, the University of Manchester Institute of Science and Technology (UMIST) and also the Production Engineering Research Association (PERA).

One of the early examples of GT was at Audley Valves Ltd (later Serck-Audco). In 1965 Gordon Ranson became the managing director of the company. The company was organized in a traditional functional layout. The company was experiencing a number of problems, high WIP, long lead times, etc. Perhaps the most serious problem was material handling: one component travelled over 10 km around the factory. He therefore decided to reorganize the factory into a number of flowlines, thus obtaining some of the benefits of mass production. Different products were manufactured on each of the flowlines, but these were all similar from a manufacturing point of view. This approach embodies the main concepts of group technology. Group technology is defined as follows:

> *Group technology is a technique for identifying and bringing together related or similar components in a production process in order to take advantage of their similarities by making use of, for example, the inherent economies of flow-production methods.*

The concept of flowlines is illustrated in Figure 1.14.

1.11.3 Development of GT

The concept of GT is very simple, but implementation is rather complex. Two areas need to be addressed:

Grouping components. Deriving product families to be manufactured on the flowlines is essential. A number of classification systems were developed to accomplish this task. The best known of these are Brisch and Opitz. All of the methods depend on giving a particular meaning to a particular digit in a code. The limitation of this method is that it focuses on the geometry of the part. In some cases (e.g. assembly work), the geometry is of secondary importance when devising product families. In this case, production flow analysis can be used. This uses the process routings and computer-based techniques to establish the GT flowlines.

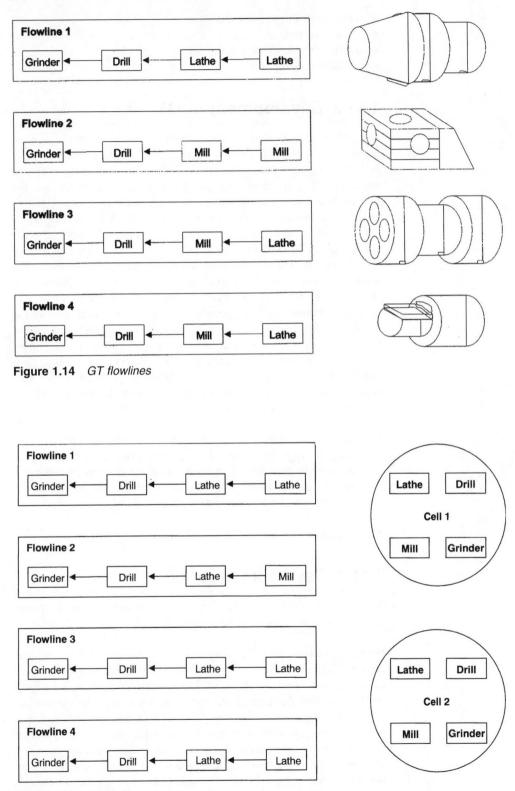

Figure 1.14   *GT flowlines*

Figure 1.15   *GT cells*

GT cells. It is often the case that some components cannot be manufactured using the GT flowlines. This problem can be solved by the introduction of GT cells (see Figure 1.15). These are small groups of dissimilar machines that are used to make those components which are not viable for flowline production. In order to determine which machines will make up the cells, it is necessary to draw up a product/process matrix.

A product/process matrix defines which products are manufactured on which workcentres. An example of a product/process matrix is shown in Figure 1.16.

While there is no apparent pattern shown in Figure 1.16, if the product/process matrix is analysed, the diagram shown in Figure 1.17 can be produced.

The outcome of the analysis is that four well-defined groups (A, B, C and D) are revealed. It would be logical therefore to create cells accordingly. It should be noted, however, that six products (9, 10, 15, 16, 19 and 22) do not fit into any of the cells.

1.11.4 Prerequisites for GT

In the 1970s, GT was promoted enthusiastically. While it is unquestionably a powerful technique, it is not applicable to every business. The following prerequisites are necessary for GT to succeed:

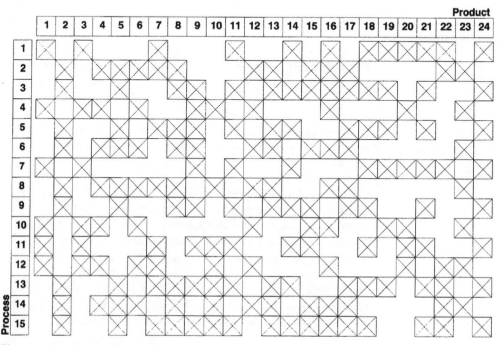

Figure 1.16   *Product/process matrix*

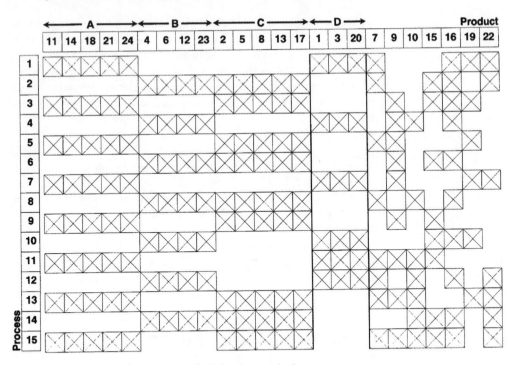

Figure 1.17 *Product/process matrix following analysis*

Natural product groups. This will allow GT flowlines to be formed.

Light components. These are suitable for movement by conveyor, etc.

Few indivisible processes. If, for example, the company has one large, expensive heat treatment plant that occurs half way through the process, this will make the formation of flowlines difficult.

A stable product mix. If plant is arranged in accordance with GT principles and the product mix changes, this could mean the layout is invalidated.

Cross skilled personnel/excess machine capacity. The GT cells need to make all of the components not manufactured on the flowlines. This means that a balanced load will not always be available for the cells. On a particular day, there may be an excess of turning work, but no milling. People will need to move to where they are required.

Willingness to undertake variety reduction. Excessive variety can render GT operation impossible. There needs to be a commitment to variety reduction by design and manufacturing engineering.

Availability of funds. Moving machines can be extremely expensive.

1.12 Flow manufacturing

1.12.1 Overview

In flow manufacturing, the path of material through the system is defined by the layout, rather than a process route. Flowline manufacture has a number of advantages in terms of throughput rates, cost and levels of WIP. Because processes are intimately connected in this mode of manufacture, it is important to minimize waiting time. This gives rise to the so-called line balancing problem. It can be applied to any circumstance when there is a limited resource, though usually this constraint is labour. There are two well-known methods: largest candidate and the ranked positional weight (RPW) method.

1.12.2 Basic definitions

The starting point for the analysis is to divide the manufacturing sequence into a number of discrete steps. These are called the minimal rational work elements and are shown diagrammatically in Figure 1.18.

This can also be expressed as shown in Table 1.1.

Each time element i is denoted te_i. Thus the total work content T_{wc} is defined as follows:

$$T_{wc} = \sum_{i=n}^{i=1} te_i \tag{1.2}$$

In the example above, T_{wc} is 48.5 minutes. How many operators are required to manufacture at the required rate and how should these people be deployed to minimize waiting time? The balance delay percentage (d) is defined as follows:

$$d = \left[\frac{nT_c - T_{wc}}{nT_c} \right] \times 100\% \tag{1.3}$$

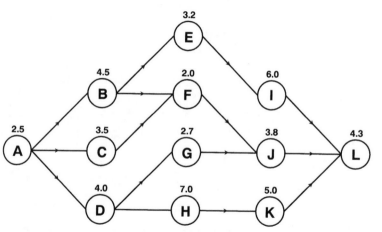

Figure 1.18
Precedence diagram

Table 1.1 Precedence data in tabular form

Element (min)	Time	Preceded by
A	2.5	
B	4.5	A
C	3.5	A
D	4.0	A
E	3.2	B
F	2.0	B, C
G	2.7	D
H	7.0	D
I	6.0	E
J	3.8	F, G
K	5.0	H
L	4.3	I, J, K

where T_c is the required cycle time for the system as a whole: i.e. the rate at which the system is required to manufacture product. [If, in the example shown in Table 1.1, the required cycle time is one product every 10 minutes and n is the number of workcentres.] It is the objective of the line balancing problem to minimize the balance delay d. There are a number of algorithms that can be applied to this problem. In complex examples, only computer-based solutions are practical. There are, however, two well-known manual approaches that can be applied. These will be discussed in turn.

1.12.3 Largest candidate approach

This is a simple technique with four steps:

1 List all of the elements in descending time order, as shown in Table 1.2:

Table 1.2 Largest candidate method – sorted elements

Element	Time (min)	Preceded by
H	7.0	D
I	6.0	E
K	5.0	H
B	4.5	A
L	4.3	I, J, K
D	4.0	A
J	3.8	F, G
C	3.5	A
E	3.2	B
G	2.7	D
A	2.5	
F	2.0	B, C

2 Starting at the top, select the first feasible element (note that this should not be preceded by any other task). For each workstation, the sum of the time elements should not exceed the target cycle time (10 minutes). Note also that precedence of elements should be maintained.

3 Repeat step 2 until no further elements can be added without exceeding the target cycle time.

4 Repeat steps 3 and 4 until all the elements have been utilized.

In Table 1.3, the first choice must be A, as this is forced by precedence. The next element to be chosen could be B, C or D. B is chosen because it appears highest on the list in Table 1.2. After allocating B, precedence would allow a choice of C, D or E (note that this is because E is preceded by B). All of these, however, would lead to a sum at the workcentre of greater than 10 minutes (which is not permitted). All that can be done therefore is to start on another workcentre. D is selected initially because it appears highest on the list in Table 1.2.

Table 1.3 Largest candidate method – completed allocation

Workcentre	Element	Time (min)	Sum at W/C
1	A	2.5	
	B	4.5	7.0
2	D	4.0	
	C	3.5	
	F	2.0	9.5
3	H	7.0	
	G	2.7	9.7
4	K	5.0	
	J	3.8	8.8
5	E	3.2	
	I	6.0	9.2
6	L	4.3	4.3
			48.5

Calculating the balance delay (sometimes referred to as balance loss), d:

$$d = \left[\frac{nT_c - T_{wc}}{nT_c} \right] \times 100\%$$

where $n = 6$, $T_c = 10$ and $T_{wc} = 48.5$.

$$d = \left[\frac{(6 \times 10) - 48.5}{(6 \times 10)} \right] \times 100\% = 19\%$$

1.12.4 Ranked positional weight approach

This method involves calculating a ranked positional weight (RPW) value for each element. This includes all of the 'downstream' elements: thus element A is 48.5 because it is the first element. The method has three stages:

1 Calculate the RPW value for each item. The RPW value for A is 48.5 minutes because it is the start of the network. L, on the hand, is 4.3 minutes because it is at the end of the network and there are no downstream elements at all. B is a more complex example; here the downstream elements need to found by tracing from B to L in Figure 1.18. The RPW value for B is therefore B + E + F + I + J + L (4.5 + 3.2 + 2.0 + 6.0 + 3.8 + 4.3 = 23.8).

2 Sort the list into descending RPW value order (see Table 1.4):

Table 1.4   Ranked positional weight method – sorted elements

Element	RPW	Time (min)	Preceded by
A	48.5	2.5	D
D	26.8	4.0	E
B	23.8	4.5	H
H	16.3	7.0	A
C	13.6	3.5	I, J, K
E	13.5	3.2	A
G	10.8	2.7	F, G
I	10.3	6.0	A
F	10.1	2.0	B
K	9.3	5.0	D
J	8.1	3.8	
L	4.3	4.3	B, C

Table 1.5   Ranked positional weight method – completed allocation

Workcentre	Element	Time (min)	Sum at W/C
1	A	2.5	
	D	4.0	
	B	3.5	10.0
2	B	4.5	
	E	3.2	
	F	2.0	9.7
3	H	7.0	
	G	2.7	9.7
4	I	6.0	
	J	3.8	9.8
5	K	5.0	
	L	4.3	9.3
			48.5

3 Assign each element to workcentres as with the largest candidate method (see Table 1.5). In this case the balance delay will be different as one workcentre less is required ($n = 5$) as the system is better balanced:

$$d = \left[\frac{(5 \times 10) - 48.5}{(5 \times 10)} \right] \times 100\% = 3\%$$

1.12.5 General principles

The problem outlined above is relatively rare. Indeed it could be argued that a better approach would be to define convenient workcentres and then modify the processes until the line is balanced. There are, however, some general principles that are important:

Production rates and times. The capacity of a workcentre may be expressed as a unit production time or a rate. Production times are measured as time/unit (e.g. minutes per units or seconds per unit). Production rates are measured as units/time (e.g. units per minute or units per hour). Thus:

$$\text{Unit production time} = \frac{1}{\text{Production rate}} \qquad (1.4)$$

Flow manufacture. In flow manufacture (i.e. where material moves from one process to another sequentially), the system as a whole is paced by the slowest process (bottleneck). In the case of Figure 1.19, the aggregate unit production time is 6 minutes per unit.

Figure 1.19
Production line with sequential processes

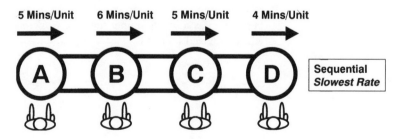

5 Mins/Unit   6 Mins/Unit   5 Mins/Unit   4 Mins/Unit

A   B   C   D   Sequential Slowest Rate

Shared resource. Figure 1.20 shows a production line with two workcentres, B and C. These are served by one operator (a shared resource). The composite unit production time is equal to the sum of the unit production times of the individual workcentres. In the case of Figure 1.20, the aggregate unit production time is 10 minutes per unit.

Figure 1.20
*Production line with
shared operator*

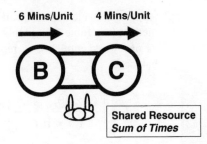

Parallel work centres. Figure 1.21 shows two parallel, identical workcentres, each with a dedicated operator. Here, the composite production rate is equal to the sum of the production rates of the individual workcentres. In the case of Figure 1.21, the composite production rate is 12 units per minute.

Figure 1.21
*Production line with
parallel workcentres*

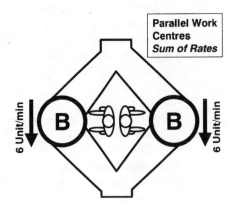

In practice, perfectly balancing a line is very difficult and requires regular review. In general, the utilization of a workcentre is defined by the system as a whole, rather than the workcentre itself. This principle will be revised in 19.11.

2 Financial accounting

2.1 Overview

The concept of accounting dates back to the fifteenth century when Paciolo developed double entry bookkeeping. Now, all companies are obliged to provide financial accounts as a legal requirement.

This chapter will examine the fundamental principles of accounting. In particular, the key financial statements are discussed. These lead naturally into the performance ratios that can be used to assess and compare companies. These elements can be described as financial accounting; that is to say accounts that provide an aggregate view of the organisation.*

2.2 General accounting principles

Accounts allow the stakeholders (owners, employees and customers) in a company to understand something of its operation. Financial accounts are targeted largely at the organization's owners and attempt to answer the question: 'Is money invested being used wisely?' The government also have a strong interest in the financial accounts as these allow appropriate taxes to be levied. It is for these reasons that keeping proper accounts is a legal requirement and in the UK, the 1985 Companies Act is the key piece of legislation defining their presentation. Because of a number of well-publicized and controversial cases of alleged fraud and changes in the commercial environment, new Financial Reporting Standards (FRS) have been developed.

Accountancy, like engineering, is a profession. This section cannot explain the broad scope of accounting and aims only to give an overview of the key concepts. It is important, however, that engineers

* Further information on accounting concepts can be found in *Business Skills for Engineers and Technologists* (ISBN 0 7506 5210 1), part of the IIE Text Book Series.

have an appreciation of accounting concepts at the very least. It should also be recognized, that all engineers wanting to reach a senior position in management must be comfortable with financial issues in general.

The guiding principles of accounting are as follows:

- **Separate entry.** In order to measure a business (and to prevent fraud), its accounts must be separate from all other entities.
- **Money measurement.** Accounts are expressed in monetary terms. Concepts such as knowledge or goodwill are not included until someone is prepared to pay for them.
- **Double entry book keeping.** All financial transactions are recorded twice – once for the giving and once for the taking of cash.
- **Realization.** If a credit sale transaction is made, it is the point of physical delivery that is the basis for reporting, not the payment.
- **Accrual.** if a credit purchase transaction is made, it is the point of physical receipt that is the basis for reporting, not the payment. This allows *matching*.
- **Continuity.** Businesses are assumed to exist as going concerns. Valuation does not assume disposal.
- **Stability of money values.** Inflation is not considered during the preparation of accounts.

2.3 Single and double entry accounting

2.3.1 Single entry system

The simplest form of accounting is single entry. This is applicable to small concerns and is commonly applied in private clubs and similar organizations. This is illustrated in the simple example of an amateur football team shown in Table 2.1.

Table 2.1   Single entry bookkeeping

Cash receipts		Cash payments	
Opening cash	£210.00	Player's shirts	£385.00
Player's subscriptions	£500.00	Ground rent	£800.00
Match tickets	£775.00	Closing cash	£300.00
	£1,485.00		£1,485.00

All of the cash flowing in and out of the club is recorded. While this approach is adequate in simple cases, it is not suitable for more complex situations. First, it does not allow for credit transactions, which are the normal way of conducting business in commercial organizations. Second, it cannot represent the diminishing value of assets (depreciation).

2.3.2 Double entry system

Double entry bookkeeping, on the other hand, provides a more complete solution. It is based on the notion of a *transaction*. This is a change in the monetary state of the organization. A transaction can only occur between two entities; for example, a company and one of its employees in the case of a payroll transaction. To keep track of these transactions, all of the entities are given an *account*. In computerized systems, these accounts are given codes (normally called general ledger (GL) or nominal ledger codes). Each transaction has two parts:

- **Debit.** This includes expenses, assets (including debt) and cash receipts.
- **Credit.** This includes income, liabilities (including debt) and cash payments.

It will be shown later that consolidation of these accounts allows the preparation of the financial statements. The double entry system is best explained by an example. Consider a company purchasing a service on credit from a particular supplier (in this example, designated 107). The work is completed on 15 November 2004 and the supplier raises an invoice for £100. This is represented by Table 2.2.

Table 2.2   Credit purchase transaction

Purchasing account – 9938		Supplier No. 107 account – 8759	
Debit	Credit	Debit	Credit
15/11/2004 Sup. (107) £100 8759			15/11/2004 Pur. £100 9938

Because of concerns about the quality of the work undertaken, the company decides to pay only half of the invoice. This takes place on 17 December 2004. This means a transfer of funds from the company's bank account to the supplier as shown in Table 2.3.

At the end of an accounting period (31 December, in this case), the transactions associated with the supplier account are summarized in the *trial balance*. This is shown in Table 2.4.

Table 2.3   Part payment

Supplier No. 107 account – 8759		Company bank account – 6757	
Debit	Credit	Debit	Credit
17/12/2004 Bank A/C £50 6757			17/12/2004 Sup. (107) £50 8759

Table 2.4   Trial balance

Supplier No 107 Account – 8759

			15/11/2004	Pur. 9938	£100
17/12/2004	Bank A/C 6757	£50	31/12/2004	BALANCE	£50

In a medium sized company (with say 800 employees), there would probably be thousands of different accounts. For these to be meaningful as a management tool, they need to be aggregated into *financial statements*. The standard reports generated are the profit and loss (P/L) account and the balance sheet. This process is shown in Figure 2.1.

Figure 2.1
Production of financial statements

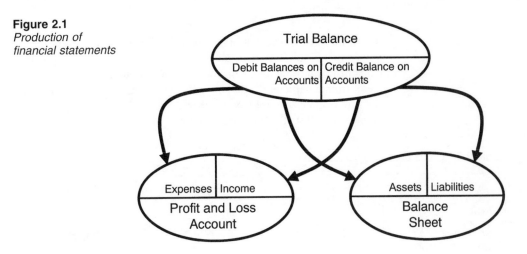

2.4 The main financial statements

2.4.1 Basic principles

The two main financial statements are the balance sheet and the P/L or trading account. The balance sheet is a snapshot of the company at a moment in time. The P/L account is a representation of activity over a period. These are not the only financial statements; for example, some companies also produce a cashflow statement. The balance sheet and P/L account, however, are universally employed. These financial statements are best understood by an example.

Imagine that a student was a company, starting in 2005. In the simplest case, a student starts with £1000 in cash (an asset). This also represents the starting capital (a liability, as someone might want this money back). For all balance sheets, the following equation is always true:

$$Assets = Liabilities \tag{2.1}$$

If the student earned £4500 in a part-time job over the year, but only spent £4300, then the profit would be £200 over the period 2005 to 2006 using the equation:

$$Profit = Income - Expenses \qquad (2.2)$$

At the end of the year, the student would have £1200. The liabilities would be the original £1000 plus the £200 profit that has *not* been given to the investors. This situation can be summarized in the financial statements shown in Table 2.5.

Table 2.5   Balance sheet and P/L accounts – simple case

Balance sheet 2005		P/L 2005–2006		Balance sheet 2006	
Assets		**Income**		**Assets**	
Cash	£1000	Earnings	£4500	Cash	£1200
		Expenses			
Liabilities		Spending	£4300	**Liabilities**	
Capital	£1000			Capital	£1000
		Profit	£200	Retained profit	£200

2.4.2 The effect of loans

Loans are not shown as income on the P/L account as they need to be repaid. They are, however, a liability. Consider the case where the student only earned £3500 and took out an interest-free loan for £1000. This would be represented as shown in Table 2.6.

Table 2.6   Balance sheet and P/L accounts – effect of loans

Balance sheet 2005		P/L 2005–2006		Balance sheet 2006	
Assets		**Income**		**Assets**	
Cash	£1000	Earnings	£3500	Cash	£1200
		Expenses			
Liabilities		Spending	£4300	**Liabilities**	
Capital	£1000			Capital	£1000
				Loan	£1000
		Profit	−£800	Retained profit	−£800

Notice now that profit is negative (i.e. a loss).

2.4.3 The effect of assets

All companies have assets. These can be of two forms:

- **Tangible.** This includes cash, buildings and stocks.
- **Intangible.** This includes goodwill, knowledge and brand names. As previously mentioned, these are normally included on a balance sheet only if their value has been quantified (i.e. by a purchaser placing a value on them). This is a controversial issue among accountants.

Tangible assets can be subdivided into fixed and non-fixed classes. Fixed assets include buildings, vehicles, plant and equipment. The non-fixed assets are cash, debtors and stock.

If the student buys food and has some remaining at the end of the period, this would be considered to be stock. This would impact the financial statements as shown in Table 2.7.

Note that stock appears as an asset on the balance sheet. This, to some extent, contradicts just-in-time (JIT) principles.

Fixed assets are also shown on the balance sheet. These, however, are subject to depreciation i.e. the value of the asset diminishes over time.

Table 2.7 Balance sheet and P/L accounts – effect of stocks

Balance sheet 2005		P/L 2005–2006		Balance sheet 2006	
Assets		**Income**		**Assets**	
Cash	£1000	Earnings	£3500	Cash	£1200
		Stock. loss/gain	£200	Stock	£200
		Expenses			
Liabilities		Spending	£4300	**Liabilities**	
Capital	£1000			Capital	£1000
				Loan	£1000
		Profit	−£600	Retained profit	−£600

Table 2.8 Balance sheet and P/L accounts – effect of depreciation

Balance sheet 2005		P/L 2005–2006		Balance sheet 2006	
Assets		**Income**		**Assets**	
Cash	£1000	Earnings	£3500	Cash	£1200
Car	£2500	Stock loss/gain	£200	Stock	£200
				Car	£2000
		Expenses			
Liabilities		Spending	£4300	**Liabilities**	
Capital	£3500	Depreciation	£500	Capital	£3500
				Loan	£1000
		Profit	−£1100	Retained profit	−£1100

Different strategies are applied in different industries; a common method for manufacturing plant is straight-line depreciation over five years, where the plant value is reduced by 20% each year down to zero. In this way, the impact of investment on profit is spread over a period of time, rather than being concentrated in the year of purchase. If the student has a car and the above depreciation strategy was applied, this would be represented as shown in Table 2.8.

2.4.4 Performance ratios

It is difficult to compare the performance of different businesses based on the formal financial statements. A profit figure of £1 M appears impressive. If the capital base of the company is £1 Bn, however, performance is actually poor. For this reason, companies often use a series of ratios. The main ratios are as follows:

$$\text{Return on capital employed (ROCE)} =$$

$$\frac{\text{Profit}}{\text{Capital employed}} \quad \text{e.g.} = \frac{£10\,\text{M}}{£50\,\text{M}} = 20\% \tag{2.3}$$

$$\text{Return on sales (ROS)} = \frac{\text{Profit}}{\text{Sales}} \quad \text{e.g.} = \frac{£10\,\text{M}}{£80\,\text{M}} = 12.5\% \tag{2.4}$$

$$\text{Asset turns} = \frac{\text{Sales}}{\text{Capital employed}} \quad \text{e.g.} = \frac{£80\,\text{M}}{£50\,\text{M}}$$

$$= 1.6 \text{ times per year}$$

$$\therefore \text{ROCE} = \text{ROS} \times \text{Asset turns} \tag{2.5}$$

Other ratios are also used, but these are beyond the scope of this book. The ratios described above are useful for senior management, but as will be seen later, they are dangerous if used as a decision-making tool when taken in isolation.

2.4.5 Dangers and limitations

The financial systems appear completely logical and deterministic at first glance. On detailed inspection, however, it is clear that these systems have limitations. Indeed, the standard financial statements can be dangerous without adequate understanding.

First, it must be recognized that the financial statements are highly aggregated and thus hide a substantial amount of detail. It is easy to make sweeping generalizations based on the accounts and their associated performance ratios. Similarly, the accounts do not reveal the root causes of good or poor performance presented in the financial statements. The accounting systems, because they focus on numerical values, can also lead to short-termism. For example, a company might invest substantial sums in research and development. In the long term, this may lead to higher profits. In the short term, however, this will have a negative impact on the financial reports.

Paradoxically, the balance sheet and P/L account do not necessarily reveal cashflow problems, which is the commonest mode of business failure. The accounts can also lead to poor practice. For example, stock is shown as an asset on the balance sheet. Stock reduction therefore, which can lead to huge operational improvements, has minimal impact on the accounts. Finally, in practice, the financial statements are complex and open to interpretation. There have been several high-profile cases of companies that are nominally sound, but have become insolvent.

Summary

In commercial organizations, the financial accounts are highly complex and can be difficult to understand, even by professionals. Nonetheless, financial accounting is the basis for all commercial activity and is a legal requirement for all businesses. It is a tool for the management of the organization on the large scale, though its influence spreads throughout companies at all levels. For this reason, engineers and technologists need to be financially aware, especially if they have aspirations to a senior position. Having said this, the financial accounts are only one representation of a company and this may not be the most meaningful, or the best. There is at present great controversy in the accountancy profession about the nature of assets. Some companies (e.g. Microsoft and Coca Cola) have nominal valuations far in excess of their tangible assets. This has led some commentators to argue that the traditional accounting approach does not now give a fair impression of true worth. Finally, accounting systems are not, by their nature, without flaws. Financial accounts can only assist, but not replace, effective management.

3 Management accounting

3.1 Overview

Financial accounts are concerned with the operation of the business at the macro scale. They are a critical element in the strategic decision-making process. They are, however, of minimal assistance at the tactical level. Management accounts attempt to address this problem.

This chapter will explain the main concepts of management accounting. It will address the areas of costing and budgeting and an area of particular importance to engineers, investment appraisal.*

3.2 Costing

3.2.1 Costing terminology

Crucial to management accounting is the notion of cost. It is defined as follows:

The money spent in undertaking some activity.

Costs are important for several reasons. In a manufacturing environment, costs are required for the valuation of work in process (WIP) as this is a legal requirement (see Chapter 2 on financial accounting). In some companies, costing is used as the basis for setting prices and in some cases this approach is entirely legitimate. There are cases, however (as will be seen later), where this approach is flawed. The main reason for costing is to support the process of management. For example, it assists managers in deciding whether an activity is worthwhile.

* Further information on accounting concepts can be found in *Business Skills for Engineers and Technologists* (ISBN 0 7506 5210 1), part of the IIE Text Book Series.

The following terms are commonly used in this subject area:

- **Cost centre.** A cost centre is a collection of activities that can be grouped together for assessment at the management accounting level. A cost centre might by a group of machines in a factory, a sales office or a team of installation engineers. Normally, a cost centre would be under the control of a manager.
- **Cost unit.** This relates to goods or services provided by an organization. As such, these units attract costs. In a manufacturing environment, the cost unit would typically be a batch of products for a customer. In a service environment, a cost unit might be an installation project. In either case, the sophistication of costing methods reflects the complexity of goods or services provided.
- **Direct and indirect costs.** Direct costs can be readily identified and related to a cost unit. For example, the cost of raw material can be easily related to the cost of a batch of products. Costs such as lighting and heating, however, cannot easily be attributed to individual cost units. Indirect costs are often referred to as *overheads.*
- **Fixed and variable costs.** This point is related to the one above. Fixed costs are those that are incurred irrespective of activity. For example, the cost of rent and rates for a retail organization are independent of sales. Variable costs, on the other hand, vary with activity (usually linearly). For example, in a foundry, energy costs will be strongly related to the amount of material produced. This is illustrated in Figure 3.1.

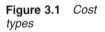

Figure 3.1   *Cost types*

The total cost of a unit is given by the following expression:

$$\text{Total Cost} = \text{Labour} + \text{Material} + \text{Expense} \tag{3.1}$$

- **Labour.** This is the cost of people associated with the unit. Some of this cost is direct (i.e. the people actually doing the work). Some labour costs are indirect, however (e.g. supervision and support staff).
- **Material.** Many goods and services require materials. Most material costs are direct, though it is convenient to consider some materials as indirect (e.g. cheap items such as glue may not be directly related to a cost unit).
- **Expense.** Expenses can also be direct or indirect. A direct expense might be the cost of a subcontractor working on a particular cost unit. In most cases, however, expenses (such as buildings, rent, light, heating, etc.) are indirect.

3.2.2 Overhead recovery

In most organizations, indirect costs represent an increasing proportion of the total. Dealing with these costs therefore is extremely important, though this can be challenging for accountants. Setting aside the crude alternative of simply splitting overhead equally between cost units, apportioning can be accomplished using three basic methods:

● **Full absorption.** In this case, the entire overhead is spread evenly across cost units in proportion to their direct costs.
● **Departmental.** Here, costs are not spread evenly. If cost units do not require certain departments, then their overhead is not included.
● **Activity based.** This is a relatively new technique. It is based on the notion that activity causes overhead cost. An activity might be 'obtaining sales orders' and this would involve a number of areas in the organization including sales, marketing and design. The costs of these areas would then be apportioned in the ratio of the number of transactions made.

Methods of overhead recovery are controversial both at an academic and individual case level. To illustrate this point, consider a college with three departments: Art and Design, Biology and Commerce and Industry. The college as a whole has overhead costs of a library (£250 K pa), the registrar's department (£100 K pa) and the buildings (£500 K pa for a total area of 2500 m^2). The simplest way to apportion the overhead would be to divide it equally among the departments. This is illustrated in Table 3.1.

Table 3.1 Simple apportioning of overhead

Scheme 1 – Crude		Direct costs						
Department	Stud. No.	Lab.	Mat.	Exp.	Dir.Tot.	O/Head	Tot.Cost	Cost/St.
Art and Drama	300	£240	£20	£15	£275	£283	£558	£1.86
Biology	180	£216	£50	£70	£336	£283	£619	£3.44
Commerce and Industry	600	£360	£5	£5	£370	£283	£653	£1.09
Total/Average	**1080**	£816	£75	£90	£981	£850	£1831	£2.13

Key: Stud. No. = No. of students   Lab. = Labour Cost   Mat. = Material Cost
      Exp. = Expenses   Dir.Tot. = Total of Direct Cost   O/Head = Overhead Cost
      Tot.Cost = Total Cost   Cost/St. = Cost per Student

Based on Table 3.1, it appears to be expensive to educate a biologist. The Art and Drama Department complains that the overhead charge of £283 K is too high, as their direct costs are relatively low. If the overhead is apportioned in the ratio of the direct costs, Table 3.2 is the result.

At this point, the Biology Department complains. This department uses very little space – one of the reasons their expenses are so high is that extensive fieldwork is required. As a result, the cost of buildings is allocated in proportion to the space each department uses (Table 3.3).

There could be further arguments: Biology could argue that because they have few students, they place relatively low demands on the

Table 3.2   Overhead apportioning relative to direct cost

Scheme 2 – Full Absorb				Direct costs				
Department	Stud. No.	Lab.	Mat.	Exp.	Dir. Tot.	O/Head	Tot.Cost	Cost/St.
Art and Drama	300	£240	£20	£15	£275	£238	£513	£1.71
Biology	180	£216	£50	£70	£336	£291	£627	£3.48
Commerce and Industry	600	£360	£5	£5	£370	£321	£691	£1.15
Total/Average	1080	£816	£75	£90	£981	£850	£1831	£2.12

Table 3.3   Overhead apportioning taking into account space used

Scheme 3 – Spc.Comp.					Direct costs				
Department	Stud. No.	Space	Lab.	Mat.	Exp.	Dir. Tot.	O/Head	Tot.Cost	Cost/St.
Art and Drama	300	800	£240	£20	£15	£275	£258	£533	£1.78
Biology	180	300	£216	£50	£70	£336	£180	£516	£2.87
Commerce and Industry	600	1400	£360	£5	£5	£370	£412	£782	£1.30
Total/Average	1080	2500	£816	£75	£90	£981	£850	£1831	£1.98

registrar's department. Art and Design could argue that their students do not use the library as much as the Commerce and Industry Department. In practice, the process of overhead allocation is arbitrary and as with financial accounting, the judgement of managers is essential to making reasonable decisions.

3.2.3 Effect of volume

Without an understanding of the nature of overheads, poor decisions can be taken. If, as in the examples, the college decides that it should close the Biology Department (as students cost too much), this will have the effect of increasing overhead in other areas. It is a common fallacy in industry that 'high cost' activities are unprofitable and should be discontinued. In some cases, this might be true. In others, discontinuing an activity can lead to massively increased costs for those that remain. It is important therefore that managers make decisions based on a reasoned overview of the situation.

3.3 Setting prices

Some companies use costs as the basis for setting prices. Typically, the following equation is used:

$$\text{Selling price} = \text{Cost} \times \left[\frac{100 + \text{Profit margin}}{100} \right] \qquad (3.2)$$

In some cases, this method of setting prices is built into the agreement between customer and supplier (as is sometimes the case in Government Contracts). In a free market environment, this approach has inherent problems. Because of the nature of overhead apportionment, it is inevitable that some activities will be costed as being 'unfairly' high and others too low. Those activities with high costs will receive a high price and might not be competitive. Sales volume may fall and increase overhead costs on other activities. It is recommended therefore that commercial organizations should charge on a 'what the market will stand' basis. In some environments, increases in prices will have little impact on sales revenues.

3.4 Budgeting

All organizations, except the smallest, need to go through a process of budgeting. Each manager in the organization needs to be provided with a budget so that he/she will know what resources are available to fulfil the company's objectives. Without budgeting, senior management would need to be intimately involved with every decision made in the organization. The process of budgeting can be summarized as shown in Figure 3.2.

Figure 3.2
Budgeting overview

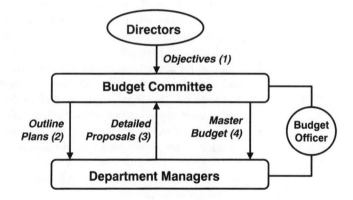

The process is an example of so-called 'top-down, bottom-up' management. Each year, the company's directors define overall objectives and these are passed to a budget committee consisting of directors and functional managers. The budget committee produces outline plans that are presented to each departmental manager who is required to respond with detailed proposals. In turn, the budget committee produces a master budget that is generally fixed for the year. To support the process, an accountant is usually nominated as the budget officer to provide information/advice.

The detail of the budgeting process is shown in Figure 3.3.

The first stage is to identify the *key factor* that constrains growth. It is from this factor that all subsequent budgeting activity will flow. In

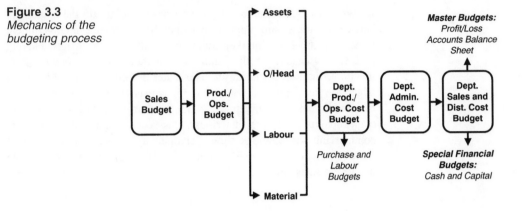

Figure 3.3
Mechanics of the budgeting process

most cases, this will be sales, though on occasion it will be space, capacity or materials. For the sake of simplicity, sales are considered to be the key factor for this section.

The sales and marketing department define projected sales volumes for the year, split into monthly subtotals. Sales will further be subdivided by product groups and customers. The next stage is for the production or operations budget to be specified. This will define how the sales budget will be met. In a manufacturing company, this will in effect define what is planned to be produced. In turn, this can be used to define what assets, materials, labour and associated overheads will arise from the production/operations budget. This information can then be broken down into budgets by individual departments. Production or operational departments will need budgets and, in particular, levels of purchasing and labour will need to be defined. Budgets will also be required for administrative departments and those associated with sales and distribution. Once these activities are complete, a master budget can be produced. This is an aggregate of all of the lower level budgets. This can be used to produce a projected P/L account and balance sheet. These documents need to be approved by the senior management team. In addition to the above, two special budgets are also produced:

- **Cash budget.** This defines the cash inflows and outflows from the business by month.
- **Capital budget.** This defines spending on capital equipment.

These two budgets are important as they define what level of borrowing will be required.

3.5 Capital investment justification

3.5.1 Overview

One area of accounting which is crucial to engineers particularly is the need to justify capital investments. That is to say, to answer the question: Is it appropriate to spend money to purchase plant and equipment? The reasons a company might purchase new equipment are as follows:

- **Potential for increased sales.** This would generate more revenue.
- **New products or service.** New opportunities for sales could be enabled.
- **Costing savings.** The new plant might be cheaper than existing equipment.
- **Replacement.** This is required if existing plant has reached the end of its useful life.
- **Change of financing.** This is the case where existing plant is rented or leased.

Four methods of investment appraisal will be discussed below.

3.5.2 Simple methods

Consider a company that is considering the purchase of a machine costing £100 K. The company concerned justifies capital expenditure over five years. The machine is projected to generate cashflows of £25 K pa once commissioned. This can be represented by Table 3.4.

Table 3.4 Payback method

Year	C/F (£K)	Cum C/F (£K)	
0	−£100	−£100	
1	£25	−£75	
2	£25	−£50	
3	£25	−£25	
4	£25	£0	Break even
5	£25	£25	

Key: C/F = Cashflow   Cum = Cumulative

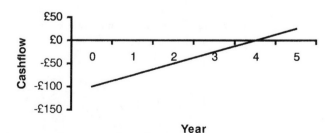

Payback Method

Figure 3.4
Payback method

Examining Table 3.4 and Figure 3.4, it is clear that the payback time is four years. This may or may not be adequate to convince the senior management that the investment is justified (in British industry, probably not). This fails to take into account the remaining cashflow in year five, however.

Another method is rate of return. This compares the average profit over the period (in this case £25 K final cumulative cash flow spread over five years) with the average investment. In the example above, let us assume plant is depreciated over ten years. This means the plant still has half its value at the end of the justification period. Thus the average investment is £50 K. This is illustrated in the following expression:

$$\text{Rate of return} = \left[\frac{\text{Average profit}}{\text{Average investment}}\right] \times 100\%$$

$$= \left[\frac{5}{50}\right] \times 100\% = 10\% \tag{3.3}$$

3.5.3 Time value of money

The previous analysis has assumed that the monetary value of different alternatives remains fixed over time. Actually, cashflows in the near future are of greater economic value than those in the distant future. This conforms to common sense; if a positive cashflow is obtained, the sooner this can be invested to advantage, the better. If a negative cashflow occurs, the later this occurs, the later interest on a supporting loan can be deferred.

Because of the above, it is necessary to consider the *time value of money*. This concept can be summarized, as money in the future is worth less than in the present.

3.5.4 Net present value

To take into account the time value of money, it is necessary to convert all cashflows to a single standard. This is the so-called *present value* (PV). The value of cash at time t is given by the following expression:

$$\text{PV} = V_t\, D_f \tag{3.4}$$

where V_t = value at time t and D_f = discount factor.

The discount factor is given by the following expression:

$$D_f = \frac{1}{\left(1 + \left(\dfrac{r}{100}\right)\right)^t} \tag{3.5}$$

where r = discount rate (%) and t = time (years).

This is best explained using an example. Consider the case in 3.5.2 with an investment of £100 K yielding cashflows of £25 K pa for five years. Because, in practice, money does not have a constant value it is necessary to discount the later cashflows. In the previous example, let us assume the company accountant set the rate at which money diminishes in value as being 15% (i.e. discount rate = 15%). This is illustrated in Table 3.5.

Table 3.5 Illustration of the time value of money

Discount rate = 15%

Year	C/F	Cum C/F	Dis.Fac	PV	Cum PV
0	−£100	−£100	1.00	−£100.00	−£100.00
1	£25	−£75	0.87	£21.74	−£78.26
2	£25	−£50	0.76	£18.90	−£59.36
3	£25	−£25	0.66	£16.44	−£42.92
4	£25	£0	0.57	£14.29	−£28.63
5	£25	£25	0.50	£12.43	−£16.20

Key:	Dis.Fac = Discount Factor	PV = Present Value	All monetary values are £K

Note that the cumulative present value at the end of the justification period (or net present value − NPV) is −£16.2 K. The accounts department would interpret this as meaning that the investment was not justified.

Another way this same technique can be applied is to calculate internal rate of return (IRR). Here the discount rate is calculated (usually by trial and error on a spreadsheet) such that the NPV is zero at the end of the period. This gives perhaps the best all-round view of the investment (see Table 3.6).

IRR has the advantage of allowing alternative investments to be compared.

Table 3.6 Table illustrating IRR

IRR = 7.93%

Year	C/F	Cum C/F	Dis.Rate	PV	Cum PV
0	−£100	−£100	1.00	−£100.00	−£100.00
1	£25	−£75	0.93	£23.16	−£76.84
2	£25	−£50	0.86	£21.46	−£55.38
3	£25	−£25	0.80	£19.88	−£35.49
4	£25	£0	0.74	£18.42	−£17.07
5	£25	£25	0.68	£17.07	£0.00

3.6 Summary

Management accounting is a tactical decision-making tool. It allows plans to be monitored and potential problems to be detected. In the case of budgeting, it also allows work to be delegated by senior management. In the case of justification, it allows investment decisions to be rationally appraised. This is of course vital to all engineers. Like financial accounting, however, management accounting has limitations. It is based on a number of assumptions. Without understanding these assumptions, it is easy to draw erroneous conclusions. It is essential that care and judgement be exercised.

3.7 Case study

3.7.1 The sharp cookie

George Newton was confused. He couldn't understand why he should break biscuits. He owned a small bakery that made a variety of products from bread to cakes. One of his personal favourites were the biscuits, because they were his dear old grandmother's recipe. He sold some of his produce in his own shop, but others he sold to other retailers. George was not a natural businessman, but he always made money because he loved baking and it showed in the taste of his produce. He certainly wasn't an accountant, which was perhaps why he couldn't understand his shop manager, Hari Patel.

'It's quite simple', said Hari, *'biscuits sell for 60p a dozen.'*

'Right. And it costs 10p a dozen in ingredients, so that's 50p profit for every 12 biscuits sold', said George.

'No – you've forgetting the cost of baking and retailing. I estimate that's 30p a dozen. The real profit is more like 20p a dozen. And that's OK – we have demand for around 2400 per week. That's £60 profit on biscuits alone.'

'Well, they taste great – they're my grandmother's recipe.'

'Now the oven can make 600 per day – that's 3600 per week because we don't open on Sunday. But normally we make only 400 a day because there's no point in making any more because we can't sell them.'

'Now that I understand – no point in making things we can't sell.'

'Remember last week when we dropped that big box on the floor and it broke all the biscuits.'

George frowned. It was a tragedy to break top-quality biscuits. He couldn't blame anyone though because he was the one who'd dropped them.

'I had the idea to sell them off cheap as broken biscuits', said Hari. 'We sold them by the kilo and I priced them at an equivalent price of 30 p a dozen. They sold out really quickly. The reason is that lots of people love your biscuits but can't afford 60 p a dozen. They'll accept that the biscuits are broken if they haven't got the money for the unbroken kind. So then I realized we could make money if we deliberately broke them.'

'Aha – you haven't thought this through. I can see why it's a good idea if you break them accidentally, but it can't make sense to break them deliberately. If it costs 10 p for material and 30 p for everything else that means the cost of making a dozen biscuits is 40 p. If you sell them for 30 p dozen, that's a loss of 10 p per dozen.'

Hari was becoming exasperated. 'But it costs the same to run the oven whether you make 400 or 600. And the extra effort in the bakery to make an extra 200 biscuits is minimal. In practice the only difference between making 400 and 600 is the ingredients and that's only 10 p a dozen. The profit on a dozen broken biscuits is 30 p minus the ingredients of 10 p. What we should do is sell as many unbroken biscuits as we can, then break any that are left and sell them off cheap.'

George thought hard. What Hari was saying seemed to make sense. But it was sacrilege to break his top-quality biscuits. Still, business was business. 'So what you're saying is that we make extra money on broken biscuits. You really are a sharp cookie, Hari. But could you go over the part again where we make money by breaking perfectly good biscuits?'

3.7.2 Analysis

Hari Patel is broadly correct. Intuitively, it does not seem logical to deliberately break the biscuits. But because most of the costs in the bakery are fixed, it is sensible to fully utilize capacity. The company does need to exercise caution, however; if the availability of broken biscuits starts to drive down the demand for the unbroken type, then the bakery will be worse off.

4 Management of design

4.1 Overview

In industry, new product development (NPD) is crucial to survival. NPD is the process of introducing new products for the purpose of generating sales revenue. The process of design itself is also critical. Typically, 80% of a product's cost is fixed during the design process. A schematic representation of the NPD process is shown in Figure 4.1.

There are three fundamental approaches to the NPD process:

- **Market-pull.** Products are developed in response to the perceived needs of the marketplace.
- **Technology-push.** Here, NPD is initiated because new technology is available to allow a product to be introduced.

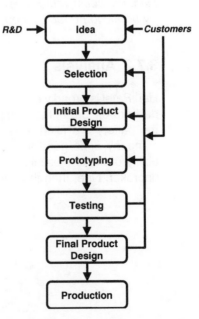

Figure 4.1 *The NPD process*

- **Inter-functional.** This is a hybrid approach that incorporates both of the points above. This is a common model for NPD.

The first stage is the generation of an idea. In principle, product innovation may arise from any source. Some ideas arise from individuals. This mode of innovation is relatively rare, despite the existence of well-known inventors such as Thomas Edison or James Dyson. Typically, ideas will arise from existing products, competitors' offerings or surveys of the customer base.

Often, a variety of ideas will be generated and the next stage is to select those that will go forward. Without this stage, NPD will lack focus, with too many projects in progress. This will be followed by initial product design. This requires significant engineering effort and is sufficiently detailed to allow a prototype to be manufactured. Following testing, final product design can be undertaken. In practice, this process is iterative and some stages will need to be repeated (at least in part). Finally, the product can go into production.

The cost of designing an advanced product, such as a commercial airliner or an automobile, is formidable. While the details of NPD will vary from organization to organization, in almost every case costs increase as the process proceeds (see Figure 4.2). The reasons for this

Figure 4.2 *NPD costs*

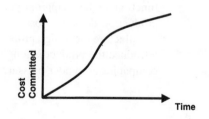

are clear; generation of ideas and selection require relatively little effort compared to detailed design or construction of a prototype. For this reason, there are great benefits in ensuring the early stages of the process are effective and that the number of changes at the later stages are minimized. A systematic process of design is therefore desirable to trap mistakes at an early stage. For this reason, there are techniques, such as quality function deployment (QFD) and failure mode effect analysis (FMEA), that aim to make the design process more formal.

This chapter will review some techniques that can be applied to the management of design.

4.2 Sequential engineering

The traditional model for NPD is sequential engineering. This is illustrated in Figure 4.3.

In this model, the various functions involved in NPD operate sequentially. For example, marketing will define a product concept

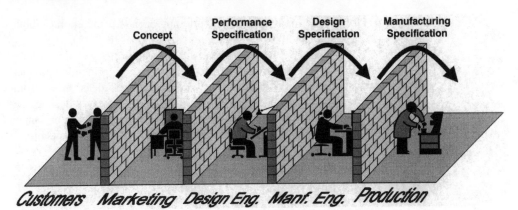

Figure 4.3   *The sequential engineering model*

based on contact with customers resulting in the definition of a performance specification. Design engineering will then convert this into a detailed design specification. Manufacturing engineering will then convert this into a manufacturing specification. This will be passed to production who will be required to make the product.

There are two main difficulties with this approach. First, there are many opportunities for the information being passed from function to function to be inappropriate. For example, it may be the case that production does not possess equipment that is sufficiently accurate to manufacture the design proposed. Second, the length of time required to introduce the product is long. As was mentioned in the foreword, most companies are under pressure to reduce NPD lead times.

4.3 Concurrent engineering

The modern approach to design utilizes the concurrent engineering approach (also called parallel or simultaneous engineering). There is no universal definition of concurrent engineering, but the basic theme is that the functions involved in NPD work together. The use of multi-functional teams can prevent problems such as those alluded to earlier. With concurrent engineering, for example, limitations in machine accuracy would be incorporated into the process *before* detailed manufacturing specifications had been finalized (thus saving considerable effort). Another benefit is that the NPD lead time can be significantly compressed. Finally, because the marketing function is involved at all times, the so-called 'voice of the customer' (VOC) is present at all stages in the process.

The concept of working in parallel has also been extended beyond company boundaries. In the automotive industry, it is now common for original equipment manufacturers (OEMs) to involve component producers at an early stage in the development of a new model.

There can, however, be practical difficulties in the concurrent approach. Communication between the members of the multi-functional teams needs to be effective. In practice, formal methods of design management are highly advantageous in a concurrent environment. This is one of the reasons for the development of quality function deployment (QFD).

4.4 Quality function deployment

4.4.1 QFD overview

One way in which structure can be imposed onto the design process is QFD. This technique was first introduced by Professor Yoji Akao and Professor Shigea Mizuno in the 1960s at the Mitsubishi shipyard in Kobe. QFD is based around a series of standard matrices. The most commonly applied is the house of quality (HOQ – so called because of its superficial resemblance to a house).

At a fundamental level, the HOQ relates the requirements of a design to the means by which those requirements will be met. In this way, it allows priorities to be defined for applying engineering effort.

4.5 QFD structure

At the simplest level, the HOQ is as shown in Figure 4.4. A matrix defines the relationship between the requirements of the design (stated from the customer's point of view) and the product characteristic (in engineering terms). The strength of the relationship is defined in subjective terms on a simple scale (this varies in different applications of QFD).

In order to understand how the HOQ works, consider the example of a drinks mug. This is illustrated in Figure 4.5. One key requirement from a user of a drinks mug is that the liquid is kept hot. This requires the mug material to be a good thermal insulator and to a lesser extent,

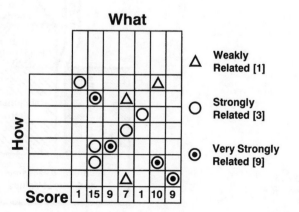

Figure 4.4   *Simple HOQ matrix*

Figure 4.5
*Example HOQ
matrix*

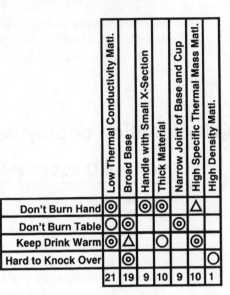

	Low Thermal Conductivity Matl.	Broad Base	Handle with Small X-Section	Thick Material	Narrow Joint of Base and Cup	High Specific Thermal Mass Matl.	High Density Matl.
Don't Burn Hand	◎		◎	◎		△	
Don't Burn Table	○	◎			◎		
Keep Drink Warm	◎	△			○	◎	
Hard to Knock Over		◎					○
	21	19	9	10	9	10	1

for the mug to have a high thermal specific heat capacity. In addition, the mug should not burn the user's hand. This requires that the mug has a small cross-section handle. It also requires that the mug material is a good insulator.

Note that the relationships need to be evaluated subjectively. Thus the HOQ is a tool for the user to provide a structured approach to design.

4.5.1 Detailed HOQ matrix

The matrices presented so far are simplified. They relate customer requirements to product characteristics. In practice, the HOQ matrix is more detailed (as shown in Figure 4.6). The core of the matrix is the

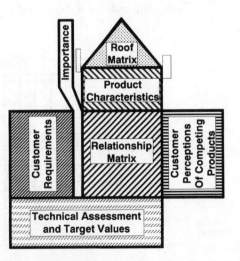

Figure 4.6
*Detailed HOQ
matrix*

same as that discussed above. There are, however, some key additions:

Customer perceptions. This box indicates customer perceptions on a scale 1–5 (5 = highly positive perception). Perceptions of competitors are superimposed on the chart.

Room matrix. This indicates the relationship between different product design characteristics. These can be positively or negatively related.

Importance. This is a value that is subjectively allocated to customer requirements on a scale 1–9 (9 most important).

Technical assessment and target values. This section provides technical information relating to the product characteristics. This section also provides target values for the product characteristics as a focus for further improvement.

An example of a HOQ matrix for a disc brake pad (DBP) is shown in Figure 4.7.

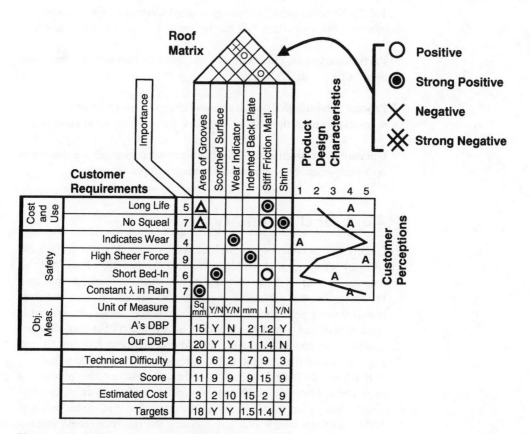

Figure 4.7 *HOQ example for a disc brake pad*

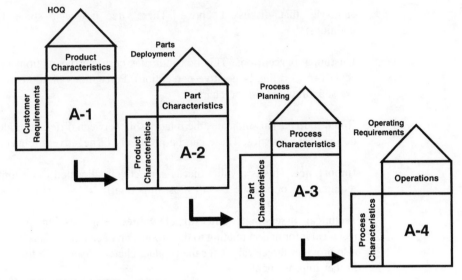

Figure 4.8 *Linked QFD matrices*

4.5.2 Extensions to the HOQ

The HOQ is the most commonly applied element of QFD. There are, however, other matrices that can be applied as shown in Figure 4.8.

Parts deployment. This examines which component parts are important in meeting the design goals.

Process planning. This examines changes to processes that are necessary to meet the revised needs of the components to be made.

Operating requirements. This examines what changes are needed in operating practices to support the changes in processes.

4.5.3 QFD summary

QFD is now in common use; indeed, it is now employed in service and software design. QFD yields a number of benefits. Principally, it encourages a better understanding of customer requirements and their influence on the technical elements of design. It also involves manufacturing personnel at an early stage in the design process and in general provides a tool for breaking down barriers between different functions in the organization. In addition to these benefits, it also provides useful data for the introduction of other new products.

Because of QFD's formal structure, it reduces the number of engineering changes (which can substantially reduce costs, particularly if late changes are avoided). Similarly, it can assist in the reduction of NPD lead-time, which is a key element for competitiveness in most businesses.

4.6 Failure mode effect analysis (FMEA)

4.6.1 FMEA overview

FMEA is a systematic technique for assessing the effects of product or process failure (note that failure means 'non-compliance with customer specification' in this context). It dates back to the late 1940s when it was developed by the US military. It assesses four factors:

- **Potential failure mode.** How can the product or process fail to meet specification?
- **Potential failure cause.** What could cause the potential failure mode to occur?
- **Current controls.** What measures are currently in place to prevent the failure occurring?
- **Occurrence, severity and detection.** These factors (assessed on a scale of 1–10) are assessed as shown in Figure 4.9. By multiplying these three factors the risk priority number (RPN) is determined. This provides an indication of the severity of problems and thus can assist in prioritizing engineering effort.

Figure 4.9
Occurrence, severity and detection

Rating	Occurrence	Severity	Detection
1	Almost Never	Hardly Noticeable	Absolutely Obvious
	Occasionally	Dissatisfaction	Visible, But Could Be Missed
10	Often	Serious Effects	Undetectable

4.6.2 FMEA example

Consider the example of a coat hook. There are two potential problems with this product. First, the hook might come off the wall, damaging plaster and causing coats to get dirty. Second, if the hook has burrs, this could snag a coat's material. These problems can be related to their potential causes (see Figure 4.10).

Potential Failure Mode	Potential Effects of Failure	Potential Failure Causes	Existing Conditions			
			Current Controls	O	S	D	RPN
Hook Falls Off Wall	Coat Dirty and Plaster Damage	Screw too Small	Instructions on Pack	4	9	6	216
		Screw Omitted	Customer Checks	1	9	2	18
Coat Matl. Snagged	Coat Gets Damaged	Burr on Hook	End of Line Visual Insp.	2	6	8	96

Figure 4.10
Design FMEA

O = Occurrence, S = Severity, D = Detection

330

Potential Failure Mode	Recommended Actions to be Taken	Potential Failure Causes	New Conditions					
			New Controls	O	S	D	RPN	Engineer Responsible
Hook Falls Off Wall	Supply Two Screws Inside Pack to Suit Hook	Wrong Use of Screws	None	1	9	6	54	
								D J Petty
Coat Matl. Snagged	Apply Plastic Coating to Hook	Burr Pokes Through	None	1	6	5	30	

84

Figure 4.11 *Application of improvements*

If action was taken to resolve the problem, then reanalysis could take place as shown in Figure 4.11.

4.7 Summary

There are a number of tools available to formalize the process of design. These should not undermine the creative ability of the designer. They can, however, provide a framework to allow the design process to be approached in a systematic manner. This is crucial in large, complex design projects.

Exercises – The manufacturing enterprise

Exercises – Manufacturing systems

1 Why is a consideration of volume and variety an essential element in a manufacturing strategy? Give examples of different types of technology (e.g. flexible manufacturing systems) and their relative advantages and disadvantages.

2 What is meant by the terms:
(a) Order qualifying criteria?
(b) Order winning criteria?
How are these changing and what effect does this have on manufacturing engineers?

3 How will methods of manufacturing vary over the life cycle of a product?

4 Why must the load on a particular workcentre be less than the capacity? What is the effect of high utilization on lead time and ability of a manufacturing system to respond to demand changes?

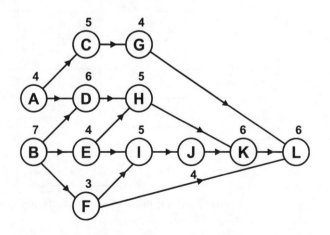

5 Assuming element times are in minutes, assign elements to workstations for a target production rate of 3 items per hour. Calculate the balance loss.

6 (a) What are GT flowlines and GT cells?
  (b) What factors are important in determining whether a particular company is suitable for group technology?
  (c) How does GT impact on plant layout and the design function?

7 What factors other than movement costs need to be considered when locating plant?

8 Where should the new plant item P (to within 1 m) be located if there are four existing plant items (see below)? The costs of transport are as follows:

  P to F1 – £0.01/metre   P to F2 – £0.02/metre
  P to F3 – £0.03/metre   P to F4 – £0.04/metre

0	1	2	3	4	5	6
1	F1				F2	
2						
3						
4					F3	
5						
6	F4					

9 Prepare a relationship chart based on the following cross chart:

Resource	1	2	3	4	5	6	7
1		40				3	3
2			20	20			
3				18	2		
4					18	20	
5						10	10
6							38
7						5	

10 What attributes of a manufacturing system might alert you to poor layout?

11 What are the different requirements for warehousing in the following types of business:
- Make to stock (MTS)?
- Make to order (MTO)?
- Assemble to order (ATO)?

Exercises – Financial accounting

1 A networking company was commissioned to undertake major installation work by a manufacturing organization on 1 January this year. The total cost of the work is £200 000. The work was completed on 1 March, though the terms of the agreement stated that only half the fee would be paid until third-party testing had been completed. £100 000 was therefore paid to the supplier on 5 March. The testing was successful and the balance of the fee was paid on 14 April. Based on the information below, list the double entry transactions and produce a trial balance for the supplier for the half-year (1 June).

Supplier name:	ANET
Supplier account code:	104781
Purchase account code:	107718
Bank account code:	105555

The first transaction is shown below:

Purch. A/C – 107718		**Supp. ANET A/C – 104781**	
Debit	*Credit*	*Debit*	*Credit*
01/03/2004 Supp. A/C ANET £200 K 104781			01/03/2004 Purch. £200 K 107718

Why was the first transaction not dated 1 January when the order was raised?

2 Consider the statements below for a company making aluminum castings:

Balance sheet 2003		**P/L 2003–2004**		**Balance sheet 2004**	
Assets (K)		Income (K)		Assets (K)	
Cash	£10	Sales	£6000	Cash	£610
Fixed assets	£3000			Fixed assets	£2400
Stock	£500	**Expenses (K)**		Stock	£500
		Spending	£4500		
Liabilities (K)		Depreciation	£600	**Liabilities (K)**	
Capital	£3510			Capital	£3510
		Profit	£900	Retained profit	£0

Calculate ROCE, ROS and asset turns based on the information above.

This information has been prepared in advance of publication for management review. Note that the profit has *not* been retained, but has been paid to the owners, reducing the cash in the business.

The operations manager has noticed that a lot of the stock is obsolete and has advanced the case that this should be scrapped. The proposed amount to be scrapped is approximately £350 K. This material could be recycled and lead to lower material purchases (though only about £10 K). Why might the finance manager be concerned about this proposal? If the proposal went through, how would this impact on the accounts?

Many football teams are now public limited companies (plcs). Most do not include the value of their playing staff on their balance sheet. What possible problems could this create?

3 A group of investment analysts are reviewing the performance of a manufacturing and distribution company. This is summarized by the graph below. They note that profits fell dramatically in the period 2004 to 2006.

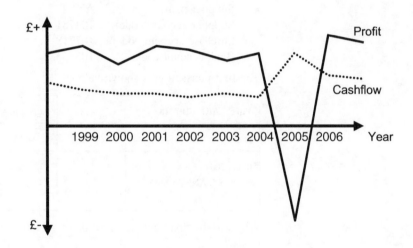

The only information that the analysts have is that a new managing director joined the company around 2005. Do you have any theories as to what might have happened?

4 Produce a set of financial statements for yourself for this year.

5 What are the limitations of financial measures for monitoring manufacturing system effectiveness?

Exercises – Management accounting

1 Four friends, Albert (A), Beatrice (B), Charles (C) and Denise (D), go out for a meal. Each has a meal (A = £19, B = £17, C = £25 and D = £19). In addition, A, B and C share two bottles of wine, costing £25

(D does not drink). They leave a tip of £25 and then B, C and D share a taxi costing £10 (A lives close to the restaurant). C pays for everything and the others tell him they will pay him their share the next day. The total cost of the evening is £140. How much should each person pay based on:
- Simple apportioning of cost?
- Full absorption?
- Departmental absorption?

2 Go Easy Jet (GEJ) Airways run transatlantic flights from London to New York via Boeing 747–400. The aircraft can carry 450 passengers and the charge is £500 per seat. The total cost of flying the aircraft empty is £150 K. The fuel cost and in-flight meal cost £50 per person flying. The finance director calculates the cost per seat as £172 500/450 = £383.33. He is happy with the profit margin of around 23%. Sometimes, however, not all of the seats can be filled. The sales director has started a scheme of selling seats to standby customers for £200. The finance director is furious as he says this will result in a loss of £183.33 per passenger. He urges the company's managing director to terminate the scheme. What is your analysis?

3 Universities could operate with highly centralized accounts systems. Expenses and costs could be controlled at the university level. In practice, however, universities undertake an annual budgeting process at department/school level, with each head having broad control over costs and expenses. What are the advantages of this approach? What would be the starting point for the budgeting process?

4 A company invests £1 M in a piece of electronics manufacturing plant. It is the policy of the company to depreciate this equipment over ten years. The project cashflows are shown in the table below.

Year	Cashflow
0	–£1 000 000
1	£300 000
2	£290 000
3	£275 000
4	£255 000
5	£230 000
6	£200 000
7	£165 000
8	£125 000
9	£80 000
10	£30 000

(a) Plot the cumulative cashflow against time and from this determine the payback period.

(b) Calculate the rate of return.

(c) Taking into account the time value of money, calculate the NPV (based on a discount rate of 12%) and the IRR.

Exercises – Management of design

1 Outline the main stages in a new product development (NPD) process. Explain why it is important that sufficient effort is expended during the early stages of the process.

2 The full house of quality (HOQ) matrix is so called because of its superficial resemblance to a house. Explain the different sections of the full HOQ matrix. Explain how the HOQ acts as a tool to facilitate communication within a multi-disciplinary NPD team.

3 A company is designing hiking boots for walking long distances over rough and hilly countryside in cold and wet conditions. Draw a simplified HOQ matrix relating customer requirements to product characteristics. Complete the matrix by making suitable assumptions about what a customer might require and the technical aspects of the design. *There is no requirement to draw the full HOQ matrix; produce a simple rectangular matrix.*

Part 2

Information Systems in Manufacturing

The system made of matchsticks

William Haycock had a bad feeling about the whole situation. He had ten years of experience working with information systems and this had all the signs of a problem. He was waiting to go into a meeting with Barry Lehman, managing director of Hartford Ltd, a manufacturing company based in the South-West of England with around 120 employees. Hartford was an autonomous part of a large engineering corporation. William was employed by the parent corporation as an information systems consultant. He had been asked to visit Hartford as they had a serious problem. Hartford made steel reinforcement units for bridges. These are complex machined items that require a substantial amount of design work. Because of the nature of the product, the process of obtaining information from customers through to eventually shipping products was complex. William had interviewed several members of staff from the company and now understood the main features of its systems. In his notes he summarized the process into five stages, A–E, as shown below:

A. A potential customer (normally a construction company) will make an initial enquiry. Hartford will provide outline information on the enquiry's technical feasibility and provide a rough estimate of cost.

B. The customer will then provide bridge drawings to Hartford who will use these to design suitable reinforcement units and provide a more precise cost quotation. This data will be incorporated into the bridge design by the customer and presented to the bridge project's commissioner (normally a local authority or the government).

C. Once the detailed plans have been checked by an independent civil engineer, the customer will advise Harford that the design has been accepted and define any minor changes that might be needed.

D. Once the commissioner has approved the project as a whole, the customer will place a firm order for the reinforcement units. At this point a final price and delivery date will be agreed and the order entered into Hartford's planning and control systems.

E. Once the products have been produced, they will be shipped to the customer.

Hartford had a computer package for managing the manufacture of the product. Most of the stages above, however, were highly specialized and complex and could not be handled by a standard package. William had discovered that one of the company's draughtsmen, Joe Davidson, had developed a PC-based system written in BASIC. This highly complex system was developed incrementally over several years and much of the work was undertaken in the draughtsman's spare time. It allowed detailed information on customer requirements to be entered and incorporated a number of intelligent features taking into account factors such as building regulations and design/costing guidelines. William had been told the package saved time for Hartford's staff as it automated many of the processes (stages A–D in William's summary).

Unfortunately, the draughtsman left the company to start a painting and decorating business. Because he spent a lot of his own time maintaining the system, faults and problems quickly started to occur once the draughtsman left. The problem had come to a head with the A821 Dunchester Bypass project. The construction company building the road needed several reinforcement units for the bridges along the route and had chosen Hartford as the supplier. Staff had been forced to return to manual methods and the resulting delays had led to angry phone calls to the managing director from the construction company. At this point, William had been called in.

One of the first things William did when he arrived at Hartford was to ask for system documentation, but the only thing that was available was a program listing (over 80 pages of BASIC code). The consultant called in the draughtsman for a day (at his standard painting and decorating rate), but the complexity of the system was such that it was impossible for him to explain how the programs worked except at the most rudimentary level.

Barry Lehman's secretary asked William to go in for his meeting. Barry stood up and shook William's hand.

'Well, what are your conclusions then?'

William knew Barry was not going to like what he was going to say and started carefully, 'I'm afraid it isn't good news. I have a number of options, but none are easy.'

'Isn't it just a matter of repairing the system faults?'

'It isn't as easy as that. The program is highly complex and its been developed incrementally.'

'What do you mean, "incrementally"?'

'Good systems are designed according to a logical process. Joe Davidson's program has just evolved. It's like he started off making a model with matchsticks without knowing what the final thing was going to look like. There's no overall plan or logical structure. In my opinion, it is unmaintainable.'

'But Joe kept the thing working.'

'I talked to him and he spent an average of 12 hours per week – his own time incidentally – maintaining the system. It was like a hobby. That amount of maintenance is unsustainable in normal circumstances. Not only that, but the system is written in an obsolete language. There isn't anyone at the corporate HQ that has skills in BASIC.'

'So what are the alternatives?'

'First, you could revert to manual systems.'

'That would cause my staff to take a lot more time in managing projects.'

'Second, we could write a new system from scratch.'

'How much would that cost?'

'£100 000, plus a maintenance fee of 15% per annum.'
'£100 000! You must be joking! How can that be right? For heaven's sake, Joe Davidson wrote the program in his spare time.'

'Joe Davidson spent hundreds of man-hours on the system and didn't spend any time on design or documentation. He just sat down and wrote code over a two-year period. Another programmer could do the same, but if that person left the company or dropped dead, you'd be right back where you started.'

Barry groaned. This was worse than he thought. 'So what's option three?'

'We could develop a simpler version of the program using a standard database management package. It wouldn't be as sophisticated and time efficient as what you've got now, but it would be fully documented and supportable over the long term. I estimate it will cost £7500. This is my recommended option.'

'So what you're saying is I've got to spend £7500 for something that isn't as good as what I've got now. That's your best recommendation?'

'I'm afraid so.' William looked around the office. He noticed that some of the wallpaper was peeling and that the colour of the paint on the ceiling was ghastly. He resisted the temptation to say that the place needed a good painter and decorator and that he knew someone who could do the job.

Analysis

Hartford had allowed themselves to become dependent on unsupportable systems. Systems had not been developed in a logical manner and eventually had failed. The company should have followed a systematic process at the outset.

Summary

This part will focus on the conceptual aspects of information systems/information technology (IS/IT) rather than detailed, technical issues of computer hardware and software. In particular, it focuses on the need for organizations to have effective information systems and how these can be designed and implemented. This part gives an overview of IT and systems in general and describes the history of IS/IT and how this is influenced by the commercial environment. It also examines the way in which organizations change and in particular how IS/IT has become a strategic issue.

Objectives

At the end of this part, the reader should:

- Appreciate how systems (in the broadest context) are relevant to modern manufacturing organizations.
- Understand how to apply system analysis techniques.
- Be able to undertake basic database design.
- Understand how IT/IS can be used to improve the effectiveness of company operation.

5 Introduction to systems analysis

5.1 Overview

The nature of the world economy is changing. There is a shift from physical manufacturing as the means of wealth creation to a knowledge led economy. For example, the contribution of manufacturing to UK gross domestic product (GDP) has reduced from more than a third to less than a quarter since 1980. Even within the manufacturing sector, it is interesting to note that UK industry now spends more money on information technology (IT) than machine tools.*

Most organizations have realized that the effective use and control of information is crucial to competitiveness. There are, however, two aspects to effective information management:

- **Technical.** This aspect focuses on the provision of the technological infrastructure to support the systems, typically concentrating on hardware and software. Because of the nature of hardware/software, development in this area is highly dynamic. To a large extent, this element is independent of organizations themselves.
- **Conceptual.** This aspect focuses on the overall design for information systems. Here, computer hardware and software themselves are less important that the organization and its requirements. Indeed, a conceptual review of a systems problem may lead to a non-computer solution. Development in this area is relatively slow compared to hardware and software. This conceptual system design can be considered as the link between the organization's information needs and the technological resources needed to support this requirement.

* Further information on IT/IS concepts can be found in *Business Skills for Engineers and Technologists* (ISBN 0 7506 5210 1), part of the IIE Text Book Series.

5.2 Evolution of IT/IS systems in business

Some of the key events in the evolution of IT/IS are as follows:

1700	Industrial revolution.
	Engineering standardization.
1800	Jacquard develops the automatic loom.
1810	Babbage's analytical engine.
1880	Hollerith perfects punch card technology and founds IBM.
1900	Fleming invents the diode.
1930	Wilson's EOQ formula.
1940	Von Neumann and Turing develop the theory of computing.
	German wartime codes cracked.
	First stored program computer developed in Manchester.
1950	Bell Laboratories develop the transistor.
	High level languages.
1960	Integrated circuits.
	On-line, real-time systems.
1970	Large-scale integration (LSI).
	Microcomputers.
1980	IBM PC launched
1900	Graphical user interfaces (GUI).
2000	Internet in common use.

IT was first introduced into business during the 1950s. The first applications were accounting oriented: payroll and ledgers. These areas were highly amenable to the *batch-processing* computer technologies then available. Because of the cost of computers, batch-processing was essential to fully utilize the hardware. Data input was, therefore, carried out off-line using media such as punched cards and paper tape. Management of computer systems was highly centralized. Indeed, Thomas Watson Sr, chairman of IBM, had remarked that the world had a requirement for a total of only five computers. During the 1960s, manufacturing applications were introduced, initially for inventory management and later in areas such as scheduling, sales order processing and purchasing. These systems were developed in isolation and were tactical in nature. That is to say, they were developed to solve specific problems. Typically, these systems simply automated existing tasks.

By the late 1960s, *real-time* systems were becoming available where users could interact with the computer. This allowed a more diverse range of applications to be developed such as computer aided design (CAD). By the 1970s, computers were common in large companies and by the late 1970s, even small companies began to employ IT, albeit in limited form. The most significant change during this period was the introduction of microcomputers. This allowed general applications such as word processing and spreadsheets to be developed. The IBM personal computer (PC) provided a standardized architecture, which led to the rapid take-up of this technology. Most significantly, this meant that

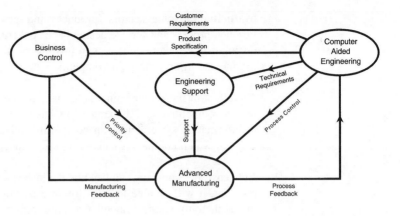

computers were available that were not under the direct control of IT professionals. This was a fundamental change; while computer science continued to evolve, this was the point at which IT ceased to be the sole preserve of a small group of technical specialists. It also meant that information systems began to be treated as an integral element of the business, in the same way as sales, marketing and financial accounting.

In manufacturing, IT is used in virtually all aspects of business operations as shown in Figure 5.1.

5.3 The commercial environment

In recent years, there have been considerable changes in the commercial environment:

Decentralization. Traditionally, companies have been highly centralized with rigid hierarchical management structures. Today, companies are typically decentralised with large organizations being divided into autonomous business units. Organizational structures have been flattened. This increases organizational agility, but also means the individual business units need to be independent.

Global markets. In the past, trade barriers protected inefficient companies. While these barriers have not been eliminated, they are now relatively weak. In addition, improvements in communications/transport infrastructure have reduced the importance of geographical location. This means that while markets are larger and there are opportunities for higher sales revenues, there is also greater exposure to competition.

Rapid change. The rate of change in the commercial environment is increasing. In some sectors, the rate at which new products need to be introduced (e.g. consumer electronics) has increased enormously. New developments, such as the World Wide Web (WWW), have radically changed the way some organizations do business.

Maturity. In some sectors, products and processes have matured. Product, process and service design have converged and there is little to distinguish different companies in this regard. This has led to the need to innovate in order to gain competitive advantage.

The effect of these four points is to make the commercial environment highly competitive and dynamic. This has led many companies to turn to information systems to gain competitive advantage. There are four ways in which companies can attempt to compete:

Low cost. If a company can become the lowest cost producer or service provider, it will be able to offer low prices. For commodity products, this is the only viable strategy. Car producers such as Hyundai, Skoda and Daewoo have adopted this approach. Companies have utilized IT/IS to reduce costs extensively. Virtually all businesses now use computerized accounting systems. These systems offer little that manual systems cannot offer. They can, however, reduce the number of personnel required.

Differentiation. In this case, the company provides a product or service that has unique features for which customers are prepared to pay premium prices (for example, BMW and Mercedes). Many companies have used the WWW in an attempt to offer customers a convenient means of gathering product information and placing orders.

Niche. Here, the company identifies a particular sector with unique requirements. Specialist car producers such as Morgan or TVR are examples of this type. Apple computers certainly occupy a market niche in providing systems for the advertising industry.

Barrier. If there is a high initial cost of entering a particular market, this is a strong disincentive to potential competitors. To some extent, all volume car producers have adopted this strategy. Car production lines and automotive design involve very high initial costs. In the computing industry, the word 'barrier' can be an emotive one. There have been a number of high-profile legal cases related to this issue.

Because of the above, IT/IS is now seen as a major competitive tool and has a high profile in many companies.

5.4 Mechanisms of meeting requirements

5.4.1 General principles

There are essentially four methods of meeting IT/IS requirements as defined by Parker. This is illustrated in Figure 5.2.

The need for a new system may arise from two sources. First, it may arise internally (i.e. by requests from system users). Second, change

Figure 5.2
*Parker's model for
system provision*

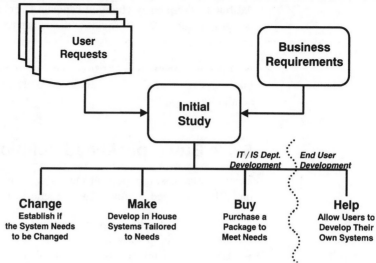

may be imposed from outside the organization (by a change in legislation for example). Based on a study of these requirements, there are four options.

5.4.2 Help – development by users

The introduction of PC-based systems has allowed users to develop their own systems. The possibilities of this approach have been expanded by the development of *client–server architectures*. This approach has three advantages. First, it ensures that users have ownership of the systems produced and their associated data. Second, it means that users do not have to communicate their requirements to an analyst. Finally, there is no need to provide external training.

There are dangers with this approach, however, and it is possible that systems are poorly designed. Pragmatically, it could be argued that this is the problem of the users undertaking the development. It is important nonetheless that the consequences of system failure are not too far reaching. The following conditions are recommended as prerequisites for user-led system development:

Single user or departmental use. Systems are not to be used widely outside the department concerned.

No direct update of corporate systems. Poor systems design must not undermine the integrity of the mainstream corporate database.

Read-only access of corporate systems. If user developed systems have read-only access to other data sources, this is relatively benign.

Management of personal data. Again, this approach is relatively benign if the data being managed are unique to the department or user concerned.

Ad-hoc or frequently changed reports. Users are often frustrated by delays in requests for information (this is one reason for the use of *data warehousing*).

No interface problems. The user interface should be simple such that the users can maintain it independently, thus ensuring system continuity.

5.4.3 Buy – packaged solutions

This is an increasingly popular method of providing systems (see 17.5 on ERP). Packaged solutions have several benefits:

- They are relatively cheap and there are no delays resulting from the need to develop software.
- Because systems are used in a number of organizations, they are subject to intensive testing.
- Systems are developed in an environment suited to holistic design.

If a packaged solution is adopted, then different solutions need to be appraised. This appraisal process has two elements. First and most important, the company supplying the software must be investigated. There are four points to consider in this regard:

Deliverables. What is actually provided by the supplier (documentation, data query language, etc.)? Most important is source code and/or a database dictionary supplied? Without these, it would prove very difficult to support the package in the case of the supplier becoming insolvent.

Usage. Is the package in widespread use? Initially, it is useful to be able to visit users in a similar business area to provide some evidence of the package's suitability.

Support. Does the supplier have the infrastructure in place to support the package? This is particularly true if an agent supplies the package. Does the vendor provide regular updates?

Organizational stability. Will the package supplier be in business for the foreseeable future? This is essential if support is to be maintained.

If the supplier is robust, then the package can be tested for its applicability to company requirements. This process can be represented by Figure 5.3.

It is extremely rare for a package and a requirement to be perfectly matched. It is essential therefore to establish the areas of functional requirement that are not covered by the package. It is important that strategies are developed to meet these requirements. There may also be areas of the package that are inappropriate to the particular requirement.

Figure 5.3
Package evaluation

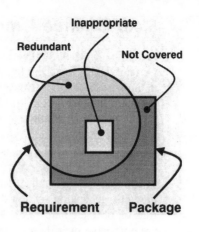

These will require a modification of the package or a change in operating practices. There are essentially three ways of using a package:

Use 'as is'. This approach is the quickest and lowest cost. With modern, highly flexible packages, this approach is growing in popularity. It allows new releases of software to be adopted easily. It does, however, require commitment from users to make the package work in the particular situation.

Modify internally. If there are problems that are insurmountable, then it may be necessary to modify the package itself. This can build knowledge internally, but does have the effect of inhibiting the take-up of new releases of the software. It is also possible to pay for these modifications from the software vendor or a third party. This can of course prove expensive.

Modify externally. An alternative approach is to develop *additional* software that exists outside the package. This is only possible if the package supports this strategy by providing appropriate functionality within the software (this functionality is sometimes called 'User Exits' or 'Hooks'). While this approach does not inhibit the take-up of new releases to the same degree as internal modifications, the problem is not eliminated entirely. One problem that sometimes arises is a dispute between customer and software vendor if an error occurs.

5.4.4 Make – bespoke development

If requirements are genuinely unique, then it is necessary to develop bespoke software. This is expensive, whether the company or a software house undertakes the development. The process can also be very slow. On occasion, business requirements evolve faster than the software and as a result, development can become very fragmented. Perhaps the most serious problem, however, is that bespoke development can institution-alize poor operating practices without great discipline. Because of the complexity of software development, it is very easy for the focus on business requirements to be lost.

5.4.5 Change – modify existing systems

Sometimes it is appropriate to modify systems that are already in place. Certainly, this can appear to be the easiest option available: it is fast and minimizes disruption to users. There are dangers with this approach, however. In practice, modification to an existing system can often mean adding new functionality on top of that which already exists. Over the long term, this can lead to overcomplex and unstructured systems. The following issues need to be addressed:

Technology. Is the application running in an environment that utilizes a previous generation system development tool or operating system, or runs on obsolete hardware? Deploying IT/IS resources on systems that are not viable over the medium term is wasteful.

Documentation. Are the existing systems that may be modified well documented? If not, modification may prove extremely difficult.

Skills. Does the organization have the skills to maintain the systems concerned? It may be the case that the systems under consideration are dissimilar to those in other parts of the organization (e.g. different hardware, operating systems, etc.). This is common when one company acquires another. It may be uneconomic to train IT/IS personnel to undertake the modification work.

Strategy. Does the modification conform to the IT/IS strategy for the organization? If it does not, it may be better to employ IT/IS resources in implementing the strategy.

Justification. Finally, is the modification justified in its own right? There is no reason why modification should not be scrutinized in the same way as other IT/IS investments.

The first three points actually indicate there is a fundamental problem that needs to be addressed irrespective of the merits of the modification. Modifications often appear attractive, but in practice can simply defer the time at which systems need to be redesigned fundamentally. Like bespoke systems, they can reinforce poor practice.

5.5 Re-evaluate requirements

5.5.1 General

One strategy that is often overlooked is to fundamentally reassess the requirements themselves. All requirements should be scrutinized to ensure that they support business objectives. Approaches such as soft systems methodology (SSM) emphasize the importance of defining the problem that needs to be addressed. Business process re-engineering

Table 5.1 Computer vs manual databases

Advantages – computer databases	Advantages – manual databases
Effort in complex systems. Simple manual systems can be created relatively easily. Complex manual systems can be very difficult to design and implement.	**Effort in simple systems.** Setting up simple manual systems can be far easier than applying information technology.
Unique reports. Computer-based systems can allow reports to be defined retrospectively. In manual systems, it is critical that reporting be considered at the point of design.	**User understanding.** Users can often find manual systems easier to understand. This is particularly true if users have limited experience of information technology.
Security (backup). Paper-based filing systems are uniquely vulnerable to fire. Copying such records is often impractical.	**Security (access).** Because information can be stored in a physically compact form, computer-based information is vulnerable to theft and/or copying.
Distributed access. Paper-based systems cannot be accessed remotely. With a basic telecommunications infrastructure, computers allow flexible data access.	**Tolerant of poor design.** In simple cases, manual systems can be modified by users *in-situ*. This reduces the importance of precise system specification.
Speed of access. Modern computer systems allow even very large databases to be interrogated very efficiently by multiple users simultaneously.	
Intolerant of poor design. Because computer database design requires great precision, this forces the system design process to be clear and well defined.	

(BPR) also encourages systems to be scrutinized more radically than simply changing software. It is also often the case that operating practices are regarded as unchangeable at too early a stage. This can inhibit the application of simple solutions to business problems.

5.5.2 Use of manual systems

With the growth of IT/IS, the use of manual, paper-based systems is often dismissed. It must be recognized that manual systems have advantages in some circumstances. Table 5.1 compares and contrasts manual and computer-based systems.

5.5.3 Symptoms of poor systems

What makes a good system is a matter of opinion. There are, however, a number of symptoms of poor system design:

Redundancy. This is where elements of the system are not used. This will usually be manifested as the printing of reports that are not used.

Duplication of effort. Different users of the system should not be undertaking the same (or very similar) tasks.

Data replication. Independent versions of the same data should not exist. This will lead to duplication of effort in maintaining two sets of records. Inconsistency can also result.

Reanalysis. It should not be necessary for users to reanalyse or reinterpret information generated by the system.

Transcription. In this case, users take output data from one part of the system and then re-enter these data into the same system.

These problems may be unavoidable. Nonetheless, the system designer should consider these problems critically. Certainly, this should trigger a careful examination of the request for a system change.

5.6 Justification for IT/IS systems

Before further discussion of IT/IS, it is worth discussing the reasons why companies need to invest in new systems at any level. These are as follows:

Initiation. If a company is starting a completely new business or extending an existing organization, there may be no alternative.

Existing capacity. If a company has expanded, then the existing system may be unable to cope with the new demands.

Cost. A new system may offer cost advantages. These may be of two types: first, the direct costs of supporting the system. Second, the operational cost of the system.

Better information. Better management information may be required, which an existing system is unable to support.

New opportunities. The development of new systems may be prompted by the availability of new technology. This may allow the company to adopt radically new methods of operation.

Forced changes. This can occur for several reasons. For example, it may be necessary if an existing system can no longer be supported. Legislation may also force changes. Finally, customers may oblige the company to undertake change.

Image. This is a legitimate reason if customer's impressions of the company are important. Simply making changes for cosmetic reasons is unwarranted.

The decision to implement a new system of any sort should always be considered carefully. System implementation is usually far more expensive that the simple direct costs (e.g. hardware and software).

5.7 System development methodologies

When considering the new system development, there is always a temptation to move directly onto programming or implementation. This often leads to poor results, particularly with complex systems. For this reason, a formal systems development methodology is often applied. This methodology is shown in Figure 5.4.

Figure 5.4 *System development*

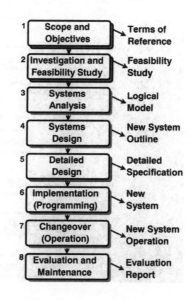

1 **Scope and objectives.** This establishes the nature of the problem. The person commissioning the new systems will produce terms of reference.
2 **Investigation and feasibility study.** This will involve an overall analysis of the operation of the existing system. This will lead to a report discussing the feasibility of implementing a new system.
3 **Systems analysis.** This requires a logical model of the existing system to be developed. Various techniques can be employed to assist in this phase. Structured systems analysis design methodology (SSADM) has a variety of tools to assist in this phase. SSADM will be introduced later in 5.10.
4 **Systems design.** In this phase, an outline design of the new system needs to be created. Again, SSADM can be used. At this stage, a realistic estimate of cost can be produced.
5 **Detailed design**. In this case, the actual operation of the system is defined. For a bespoke computer system, this will involve the

specification of programs. For a manual system, this will involve the design of forms, etc.

6 **Implementation.** In this phase, the actual work of implementation is undertaken. In the case of a bespoke computer system, this will mean programming. It will also involve documentation, procedures and training.

7 **Changeover.** It is necessary to migrate from the existing system with the minimum of disruption.

8 **Evaluation and maintenance.** This is sometimes referred to as post-implementation support. This phase is often neglected.

Few projects in manufacturing companies follow the formal method-ology. This is due in part to the fact that most projects are not too complex. It is also true, however, that many organizations do not recognize the value of a formal approach. Typically, a reduced form of

Figure 5.5
Reduced methodology

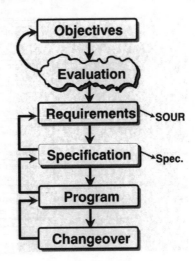

the methodology is employed (as shown in Figure 5.5). The six stages are as follows:

1 **Objectives.** It is common for the person commissioning the 'new' system not to have a clear idea of what is required.

2 **Evaluation.** It is usually a good idea for the system developer to obtain a generic understanding of the problem. It is often the case that following this stage, the analyst must re-evaluate the objectives.

3 **Requirements.** It is always worth understanding the existing system, but it may not be necessary to formally define its operation. In this case, a single document is prepared called a *statement of user requirements* (SOUR) to define what is needed from the new system.

4 **Specification.** The analyst will invariably need to define for the programming staff what is required. In some cases, this specification may be verbal. Sometimes the system developer and the programmer may be the same person.

5 **Program.** This stage is the actual coding.
6 **Changeover.** This is the process of conversion to the new system.

5.8 Importance of a structured approach

The reason for applying a formal approach to systems development is due to the fact that greater costs are expended in the later stages of an IT project than at the outset (see Figure 5.6).

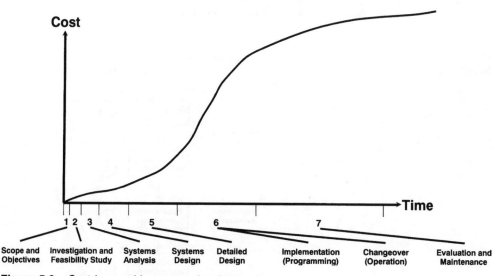

Figure 5.6   *Cost incurred in system development*

5.9 Personnel

The following terminology is often applied with respect to the people involved in the system development process:

- **User.** This is the person using a system.
- **Implementation team.** This is the team responsible for the implementation of the new system. This team usually contains system users.
- **Steering committee.** This is a group of senior managers who oversee the implementation.
- **Business analyst.** This person is responsible for defining overall requirements and design.
- **Systems analyst.** This individual is responsible for detailed system design.
- **Programmer.** This individual writes code.

5.10 SSADM overview

SSADM was developed in 1982 by the British government. The concept was that this methodology could be applied to all systems development projects for the UK government. SSADM is a structured systems analysis (SSA) methodology; that is to say, it is a predefined and formal approach to the systems design problem. Others exist, though SSADM is the most widely applied approach in the UK: indeed, 70% of British organizations, who use some form of SSA, use SSADM.

SSADM is a complex and comprehensive methodology. Because of this, this book will not attempt to address every aspect of SSADM. It is hoped to give an impression of how it can be applied and to describe some of the key features of SSADM.

Figure 5.7
SSADM flow

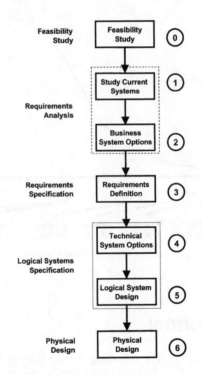

The basic flow of SSADM is shown in Figure 5.7. It has seven stages (0–6) and these are as follows:

0 **Feasibility study.** This is a preliminary stage to determine whether the proposed system is feasible before expending resources. Often, this stage is omitted, as there is commitment to the project at the outset.
1 **Study existing systems.** It is generally good practice to review existing systems whenever a new project is initiated.
2 **Business system options.** This is a broad assessment of the different strategies that could be employed to meet the requirements.

3 **Requirements specification**. Here, the requirements for the new systems are defined.
4 **Technical systems options**. This stage defines how the requirements could be met at a technical level. For example, decisions might need to be taken on what operating system or database management system (DMS) should be employed to meet the requirements.
5 **Logical design**. Here, the requirements defined at stage 3 are converted into a detailed design.
6 **Physical design.** Finally, the logical design is used as the basis for the physical development of the systems (e.g. by writing code).

To support SSADM, a number of graphical tools are available. These allow a rigorous and formal description of various aspects of the system under consideration. Some of these tools will be discussed in detail later in the book.

5.11 Systems and analysis

There are several key concepts relating to systems/analysis that need to be understood before proceeding:

Analysis. This is defined as *'Resolution into simpler elements.'* The reciprocal process is synthesis.

System boundary. This divides the system under consideration from the wider environment.

Human understanding. A large amount of research has been undertaken in the area of human cognition. This indicates that the human mind can only understand and manipulate around 7 ± 2 entities simultaneously. For this reason, it is common for concepts to be structured.

Hierarchy. Because of the nature of human thought, it is common for systems to display a *hierarchy* (see Figure 5.8). This can be seen clearly in human organizations: for example, the Military, the Church, universities and industry. This idea of hierarchy can also be observed in

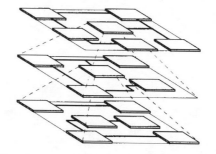

Figure 5.8
Systems hierarchy

most areas of human endeavour (except for very simple cases). For example, a car maintenance manual is separated into a number of sections (braking, fuel, power transmission, ignition, etc.), which are then further subdivided.

Functional decomposition. This is the process by which a system is broken down into logical components. With complex systems, several levels of decomposition may be necessary for a complete picture to be obtained.

Emergence. Since many systems are hierarchical, they can be considered at a number of *levels*. Properties of the system are often associated with a particular level in the hierarchy. For example, the taste of salt is *not* dependent on its constituent elements. Rather, it is dependent on the whole.

5.12 Summary

Systems are crucially important in the operation of virtually all organizations today. The design of these systems has two elements; first, the way in which the system is implemented at a physical level (hardware, program code, communications etc. Second (and equally important), the conceptual design of the system.

Like the design and manufacture of physical devices, careful planning is essential to achieving a cost-effective outcome that meets requirements. As has been shown by a number of high profile cases, the consequences of failure can be serious. One way in which the risk of failure can be minimized is the use of a formal approach to system design and the application of a formal SSA methodology.

The next chapter will investigate the early phases of the system design process.

6 System design

6.1 Overview

The early stages of a project are often the most critical. While these stages are relatively quick and inexpensive, they are often, paradoxically, ignored. One of the reasons for this is that these stages do not lead to visible progress. It is tempting therefore to press ahead with the later, higher profile activities. This chapter will examine these elements.

6.2 Terms of reference

It is often the case that terms of reference are not clearly defined by the system commissioner. It is essential that the analyst obtain a clear picture of clear requirements before starting any detailed analysis. This means that the analyst must obtain some background on the situation and possibly meet with the system commissioner to review and refine terms of reference. Definition of system boundaries is crucial to define what is *outside* the scope of the project. Without this, the project can easily lose focus.

6.3 Obtaining information and interview techniques

6.3.1 Sources of information

It is generally a good practice to understand existing systems (if they exist) before attempting something new. This requires investigation and there are a number of sources of information:

Documentation. If available (e.g. ISO9000 procedures), this can be useful. Sometimes, however, analysis of documentary sources can be tedious, irrelevant, inaccurate or worst of all misleading. Obtaining copies of forms or records can be extremely useful. Analysts will often follow particular pieces of documentation through the system to gain an objective view of its operation.

Observation. This is a very effective, if time-consuming task. One problem, however, is that people often modify their behaviour if watched.

Questionnaires. Questionnaires are useful for very large systems with many users. When used alone, they have a number of problems. These include low reply rates, ambiguous responses and misunderstanding by the respondents. Perhaps the most serious limitation is that questionnaire design is difficult without a thorough knowledge of the system.

Objective measurement. It may be possible to obtain statistical data, particularly if the existing system is computer based. A good example of the information which can be obtained is transaction rates.

Interview. This is the most common method of obtaining information. It is quick and it is possible to obtain the views and opinions of users as well as the factual operation of the system. The following points summarize good interview practice:

- Always explain the purpose of the interview.
- Always try to put the interviewee at ease. Be polite. Ask about problems.
- Pre-prepare questions. Avoid lengthy digressions.
- Never criticize the interviewee.
- Never engage in criticism of a third party or the organization in general.
- Summarize the points made by the users during the interview and also at the end.
- Take notes or use a tape recorder.
- Aim for an interview of less than 40 minutes.
- Always ask for a follow-up interview, if required.
- Return with a completed analysis after seeing other people.

Adopting the above points will usually lead to a good outcome. There are a number of pitfalls, however:

- People may not co-operate, though this is rare.
- People may answer questions in such a way as to try to please the analyst.
- People may find it hard to express themselves, particularly if the task involves judgement or the analyst is unfamiliar with the context.
- It is easy to jump to conclusions. Several people should be interviewed to obtain a full picture.

Reliance on single sources of information should be avoided. Cross-referencing of information is critical. Assumptions on the part of the analyst are dangerous. Ask additional questions if there are any doubts.

6.3.2 Reporting

At the end of a series of interviews, a summarized report of initial findings should be made. These reports should be short and should *not* use impersonal-passive style. Graphical techniques can be useful:

Organization charts. These show who works for whom in the organization. These can often be obtained from the personnel department.

Block diagrams. These show the basic flow of information in relatively unstructured form (see example in Figure 6.1).

Figure 6.1
Example of a block diagram

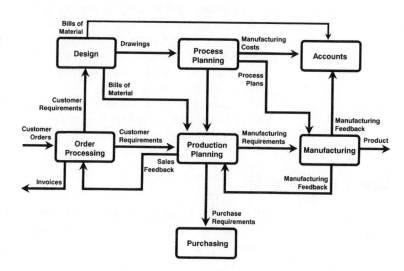

Irrespective of the style of reporting, the following areas need to be addressed:

- The individuals who perform the various system tasks.
- The information generated and how it is used.
- How is information communicated? Copies of reports and forms are useful.
- The limitations of systems.
- What alternatives exist for new methods of operation – possibly with alternatives.
- Recommendations as to how to proceed.

It should be noted that detail is important in describing the operation of a system.

6.4 Summary of initial stages

The purpose of this stage is to understand the operation of the system. It is tempting to move onto the design of new systems without first understanding the existing operation. Time spent on the existing system is usually well spent.

6.5 System design (analysis)

Following the early stages of the system development process, it is necessary to undertake systems analysis. The purpose of this process is to determine what systems are required in order to meet requirements. This process is abstract – that is to say it does not involve the creation of a physical design. For example, systems analysis does not involve the selection of hardware or detailed specification of software. In this way, decisions on these detailed (and costly) issues can be deferred.

Ideally, the system analysis process will lead to the creation of logical, formal models to represent function. One method is the creation of simple written reports. This approach has the disadvantage that written English can easily be ambiguous. For this reason, there are a number of techniques that can be employed to give a more definitive representation of a system.

This part of the book will describe flowcharts but more importantly, two of the SSADM graphical tools – resource flow diagrams (RFDs) and data flow diagrams (DFDs).

6.6 Flowcharts

The simplest technique for representing the operation of a system is flowcharting. This uses a number of standard symbols as shown in Figure 6.2.

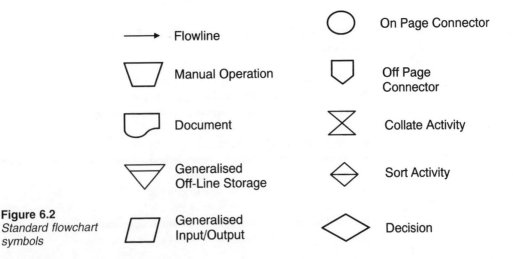

Figure 6.2
Standard flowchart symbols

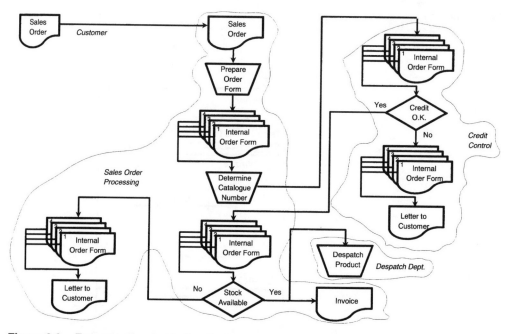

Figure 6.3   *Example of a simple flowchart*

Flowcharts have a number of advantages. First, they are simple and require little experience to obtain a good result. Second, the operation of flowcharts is intuitive and therefore the technique is easy to learn. Because of these advantages, flowcharts are easy to explain. The major disadvantage of flowcharts is that they cannot cope with realistic, complex situations. More fundamentally, the flowcharting technique does not correspond to real-world systems that display structure at a number of levels. Finally, flowcharts are not easily translated into information systems.

A simple example of a flowchart is shown in Figure 6.3.

Despite their limitations, flowcharts can still be a useful and quick way of describing systems.

6.7 Data flow diagrams

6.7.1 Overview and graphical conventions

The objective of DFDs is to define formally the relationship of activities taking place within a system and the data that are used. In order to do this, systems are represented using a small number of symbols. These are as shown in Figure 6.4:

External entity. This is something (a customer or supplier for example) that exists *outside* the system boundary. It should be noted that an external entity is not necessarily outside the organization.

Figure 6.4 *DFD symbols*

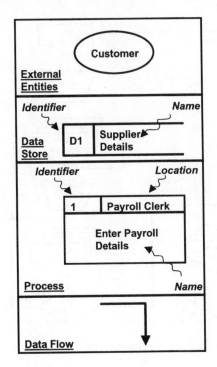

Data store. This represents a location where data are stored. It has an identifier to distinguish it from any other data store. By convention, a manual store identifier, such as a paper-based file, is prefixed by the letter 'M'. A computer-based store is prefixed by 'D'. If a store is transient (i.e. it exists only for a limited time), this is represented by the addition of the letter 'T' to the identifier. Thus the identifier for a transient manual store would be 'TM'.

Process. This transforms or manipulates data within the system. It has a numerical identifier to distinguish it from other processes. It is will be seen later that DFDs are hierarchical and this is represented in the identifier. All processes have a location. It is common for the person(s) undertaking the process to be used as the location. All processes also have a descriptive name. This must be an imperative statement (i.e. it must begin with a verb).

Data flows. Lines with arrows represent data flows. Data flows can be to or from a process. On occasion, it is also permissible for there to be data flows between external entities (though strictly speaking these are outside the system boundary). Data flows should only cross when essential.

6.7.2 Example of a DFD

Below is a simple example of a DFD. This represents a business process for paying invoices. This can be described as follows:

1 A company raises a purchase order (PO) and sends this to a supplier. This creates an entry on a computer system.

2 When the goods inwards department of the company physically receives the goods from the supplier, they create the goods received note (GRN) manually. This formally records the fact that goods have been received. The GRNs are sent to the purchasing department where they are filed.

3 When an invoice from a supplier is received, this is matched to the GRN (manual file) and the original PO (computer records). If these match (e.g. if the quantity of goods is the same on the GRN, PO and invoice), then the purchasing clerk will make an entry into a computer system to authorize payment. If the information does not match, a letter querying the invoice is sent to the supplier.

4 Once a week, a payments clerk will print cheques. This is based on the computer file of authorized invoices. As a result of this process, the supplier will be sent a cheque.

The process above can be represented more formally as the DFD shown in Figure 6.5.

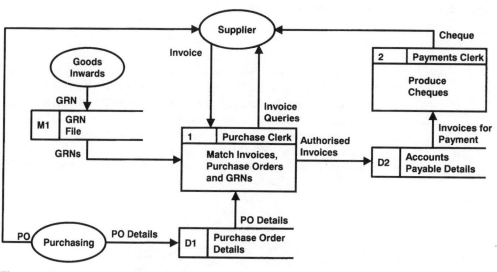

Figure 6.5 *Example of a DFD*

6.7.3 Decomposition of DFDs

In practice, systems may be far too complex to be represented on a single DFD. It is possible therefore for a process to be functionally decomposed and the system represented hierarchically. For example, in the DFD example in Figure 6.5, it is possible to functionally decompose process 1. In this process, after information matching, a transient file of query letters is created. These are sent to suppliers

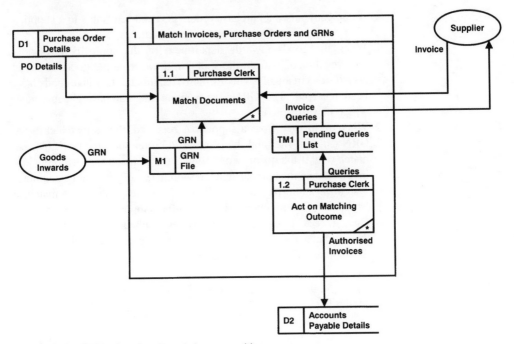

Figure 6.6 *DFD after functional decomposition*

once a week. This additional information can be represented as shown in Figure 6.6. Different levels must be consistent.

Note that the decomposed processes are given the identifiers of 1.1 and 1.2. Note that process 1.1 and 1.2 are marked with a '*' symbol. This indicates that they are bottom level processes. This nomenclature represents the hierarchical nature of the diagrams and is illustrated in Figure 6.7. Note also that different levels must be consistent.

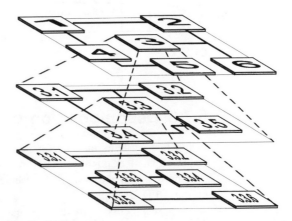

Figure 6.7 *Numbering convention for DFDs*

6.7.4 Use of DFDs

In a systems development project, four types of DFD can be created:

- **Current physical.** This is a model of the actual operations of the current system.
- **Current logical.** This explains what the systems accomplish.
- **Required logical.** This defines what the new system will be required to accomplish.
- **Required physical.** This defines how the new systems will be realized.

It should be noted that DFDs are tools for communication and in that sense are no different from natural languages. One common failing is to include large numbers of exceptions from mainstream operation. This can be confusing, particularly at an early stage. This detail must of course be dealt with at a later stage. Overly complex diagrams should also be avoided and functional decomposition employed where appropriate. It is recommended that a hierarchical approach to system design is employed, even if formal methods are not used.

6.8 Resource flow diagrams

RFDs can be used to assist in the development of DFDs in cases where a significant proportion of the existing system relates to the physical movement of items. An example of a RFD is shown in Figure 6.8. The graphical conventions are shown in Figure 6.9.

6.9 IDEF$_0$

6.9.1 Overview

Another graphical analysis technique is IDEF$_0$. This is based on a number of linked functions or tasks. These links can be of four types:

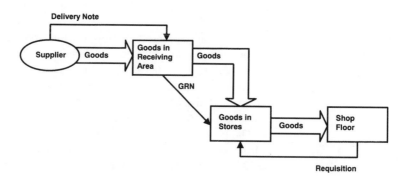

Figure 6.8
Example of an RFD

Figure 6.9 *RFD symbols*

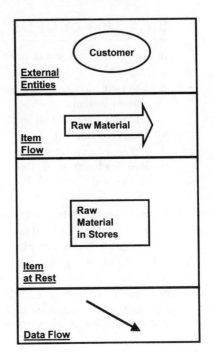

Figure 6.10 *ICOM diagram*

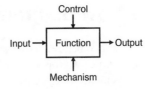

Input. If we have a task 'pay bills', then the input might be a purchase order.

Output. For the above task, the output might be a cheque.

Control. This is the 'trigger' for the task. In the above case, this might be an invoice.

Mechanism. This is the 'enabler' for the task. In the above case, this might be a computer.

The links can be represented by the mnemonic ICOM and are shown around the task in a clockwise direction as shown in Figure 6.10.

All tasks must have an input and a control. The output and mechanism are optional. Tasks are shown diagonally and linked by orthogonal lines. All links are labelled. A simple example of this technique is shown in Figure 6.11.

The power of the technique lies in the fact that any element may be further analysed and a further diagram prepared. This is shown in Figure 6.12. The system is represented by a number of levels and, in overview,

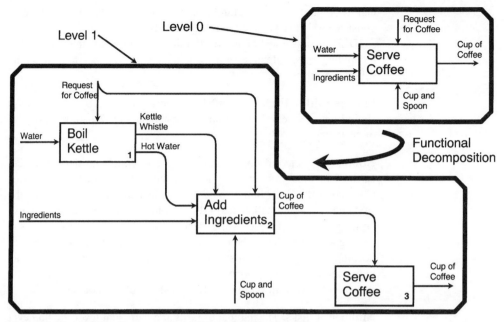

Figure 6.11   *IDEF₀ diagram – level 0 and 1 decomposition*

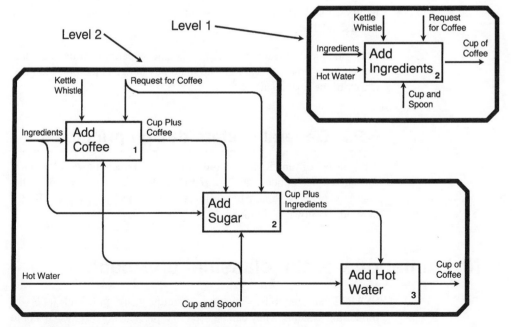

Figure 6.12   *IDEF₀ diagram – level 1 and 2 decomposition*

can be regarded as a hierarchical representation. Note that there is consistency between the hierarchical levels.

IDEF$_0$ is thus a technique for defining requirements. In order for the technique to be effective, however, it is essential to conform to the following rules:

Function titles. Each function should be represented by a *verb*. Tasks should *never* have a designation such as 'Planning Department' (this is often tempting). A better designation is 'Plan Production'.

Numbering convention. A numbering convention must be adopted. Each of the links from outside the level should be labelled – I for inputs, C for controls, O for outputs and M for mechanisms. For example, the first control should be labelled C1. Each level in the model should also be labelled. The highest level is designated (zero). Each of the tasks at the highest level should be denoted by a numeral. At the next level of decomposition, a full stop should be used as a delimiter. For example, in the simple example above the task '*Add Sugar*' would be denoted 2.2 as it is derived from the higher level task 'Add Ingredients'.

Number of tasks. The number of tasks at a particular level should never exceed eight. Again, it is tempting to ignore this rule.

The great power of IDEF$_0$ lies in the ability to check consistency. Thus, if an element has a particular number of links, there should be a correspondence at both higher and lower levels. This ability to check consistency gives this technique a great advantage over natural languages.

Linking lines should be orthogonal. When these change direction, a radius should be applied. This makes the diagrams more visibly understandable.

The principle disadvantage of IDEF$_0$ is that preparation of the diagrams can be quite time consuming. The main advantage is the precision and clarity of the final result.

6.9.2 General system design principles

The concept of functional decomposition is extremely powerful. Even if formalized techniques are not employed. It is recommended that complex systems should be designed hierarchically. A typical IDEF$_0$ diagram is shown in Figure 6.13.

6.10 Limitations of the classical approach

SSADM and the other analysis techniques are powerful tools for modelling systems. They can be used to represent existing systems or to check the validity of a proposed system. It is important, however, not to

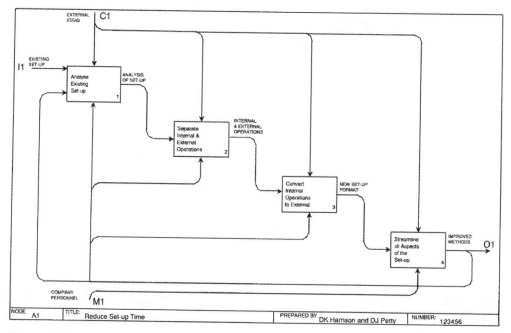

Figure 6.13   *Example of an IDEF$_0$ diagram*

simply consider the system design process as a purely mechanical process. The outcome of too many system design exercises is simply a computerized version of an existing system. There are other method-ologies for system design and implementation that exist for the sole purpose of avoiding this problem. The best known methodology of this type is soft systems methodology (SSM).

6.11 Soft systems methodology

6.11.1 Overview

By changing the fundamental operation of a system, it is often possible to make step improvements in performance. For this reason, it is entirely legitimate for the system designer to challenge the fundamental operation of a system. Indeed, some authors argue that too analytical an approach is detrimental to good system design. This approach was formalized by Checkland (1991) with the so-called SSM. This approach suggests that it is important to understand the underlying issues in a particular situation before undertaking analysis (see Figure 6.14).

The various stages in this process are as follows:

1 **The problem situation unstructured.** This is the starting point for SSM. It should be noted that the problem situation is not simply the formal systems. SSM, for example, emphasizes the importance of the people who have an interest in the problem.

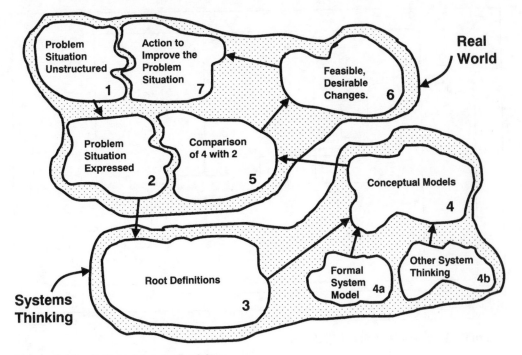

Figure 6.14   *Soft systems methodology*

2 **The problem situation expressed.** The next stage is to express the problem. Note that in SSM, the problem is not defined – this is hard systems thinking. In SSM, a so-called 'rich picture' is developed by drawing on a variety of different sources and points of view.

3 **Root definitions of relevant systems.** To provide a focus for the methodology, SSM proposed the development of a root definition. This is a formal, structured statement of a human activity system. To ensure that the root definition is comprehensive, SSM recommends that the root definition contains the following elements:

- **Customers.** These are the people who benefit from the system.
- **Actors.** The people who carry out the essential activities within the system.
- **Transformation process.** This is the process that converts inputs to outputs.
- **Weltanschaung.** This is the perspective by which the system is judged.
- **Owners.** These are people who have the power to change or terminate the system.
- **Environment.** This is the environment in which the system is located.

These elements can be remembered by the mnemonic CATWOE. A root definition for a university might be as follows:

The university is an institution for teaching, learning and research (W) responsible to the government (O) within the higher education sector (E). The activities of the institution are carried out by the university staff (A). The university's objective is to develop the abilities of its students and undertake research (T) for the benefit of the UK economy (C).

The nature of a root definition, however, is that it is subjective. For example, the following root definition for a university is equally valid to that presented previously:

The university is an institution for increasing the sum of human knowledge (W) responsible to its governors (O) within the broad context of the academic community (E). The university's objective is to provide a fertile environment for its staff and students (A) to develop their ideas and understanding (T) for the benefit of the various disciplines represented in the institution throughout the world (C).

The nature of the root definition will depend upon its author.

4 **Conceptual models.** Based on the root definition, a model is defined (normally in graphical form). This must contain the minimum necessary activities such that it can be mapped onto the root definition. Generic formal systems models have been defined to allow validation. Checkland also argues that other system thinking models should be incorporated (i.e. the ideas of other authors).
5 **Comparison of 4 with 2.** Here, the problem situation is compared with the conceptual model.
6 **Feasible, desirable changes.** These are a list of potential changes that would be beneficial.
7 **Action to improve the problem situation.** The final stage is to act on the analysis.

In practice, this methodology can be very hard to employ. There is some value, however, in producing a root definition and certainly in understanding the CATWOE elements.

7 Data analysis

7.1 Overview

Over the last 20 years, the performance of computer hardware has improved by (at least) two orders of magnitude. The performance of software has probably not improved by even a factor of ten. At the same time, in manufacturing industry it is now common to employ complex and integrated software packages to control the organization. These systems can easily control tens, or even hundreds of millions of pieces of disparate data. In practice, these elements of data are not independent, but connected together in a manner defined by the operation of the business.

These factors have led to a need to improve productivity in software development. With increasing complexity, it is essential to be able to carefully model and design information systems at a conceptual level. The purpose of this chapter is to review an important technique for the conceptual design of databases, entity-relationship (E-R) modelling.

7.2 Approaches to data storage

The traditional approach to developing computer systems was the so-called application-centred view methodology (see Figure 7.1). In this case, the logic of the programs is of paramount importance with the mechanism of data storage designed retrospectively. This approach has a number of limitations:

Redundancy. Data can be duplicated. For example, if an order processing system and credit control system have separate files, customer addresses could be recorded twice.

Inconsistency. If stock and production systems are not integrated, this can cause problems. For example, it may be impossible to reconcile WIP with the stock control.

Figure 7.1
Application-centred view

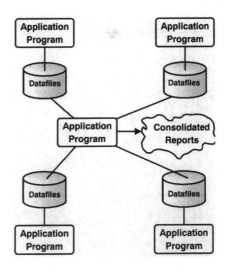

Data extraction. Writing programs to extract data from two or more programs could be difficult. If the data are inconsistent, then data extraction will be impossible.

Security. Defining access rights for different people in a multi-user environment is extremely onerous.

These limitations have led to the development of a data-centred view (see Figure 7.2). In this case, the data exist independently of the application programs. Several so-called database management systems (DMS) to facilitate this approach are commercially available. These include mini-computer-based systems such as Oracle, Ingress and Progress. PC systems are also available, most prominently Microsoft Access.

   The existence of database management systems makes the development of applications simpler. It does, however, require a formal approach to database design and structuring.

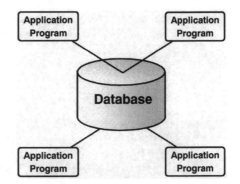

Figure 7.2   *Data-centred view*

7.3 Entity-relationship modelling

E-R modelling is a technique for formally representing data systems. Before explaining this technique, it is important to define the terminology (see also Figure 7.3):

Figure 7.3 *E-R symbols*

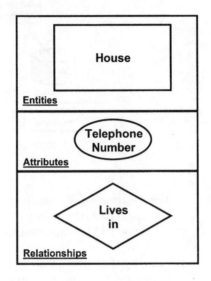

●Entity. This is an object or concept capable of independent existence that can be identified uniquely. Entities are represented by rectangles.

Attribute. This is a property of an entity. What attributes an entity possesses will depend on the context. For example, in medical records, height and weight would be a reasonable attribute of a person. In the context of a personnel system, National Insurance number would almost certainly be an attribute. Attributes are represented by ellipses. In complex situations, attributes may be omitted for reasons of clarity.

Identifier. This is an attribute or group of attributes that uniquely identifies an entity. Sometimes this is referred to as a key.

Relationship. This is an association between two entities. These are represented by diamond-shaped boxes.

These graphical conventions are illustrated in Figure 7.4.

Figure 7.4
Example of a simple E-R model

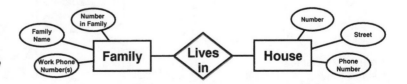

7.4 Types of E-R structure

There are a number of types of relationship between entities. These fall into a number of categories.

7.4.1 Obligatory

In this type of relationship, an entity must be connected to another entity. This type of relationship can be divided into three subclasses:

1:1. In a simple 1:1 relationship (as shown in Figure 7.5), all of the family entities are connected. The fact that this relationship is obligatory is illustrated by the fact that the 'blob' on the line is within the entity box. The obligatory 1:1 relationship is the simplest form of relationship. The implication of this in the example is that within the data set, there are no homeless families or unoccupied houses.

Figure 7.5
Obligatory 1:1 relationship

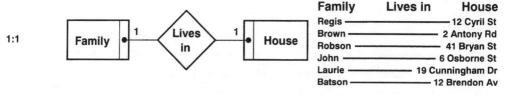

1:Many. A more complex relationship is the 1:many relationship as shown in Figure 7.6. Here, one lecturer may tutor several students, though no student has more than one tutor. The relationship is obligatory because no student is without a tutor and each lecturer tutors at least one student.

Figure 7.6
Obligatory 1:many relationship

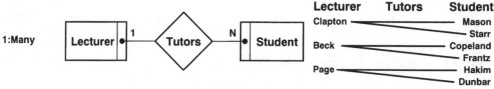

Many:many. This relationship is shown in Figure 7.7. Again, this relationship is obligatory because all modules have at least one student and all students study at least one module.

Figure 7.7
Obligatory many:many relationship

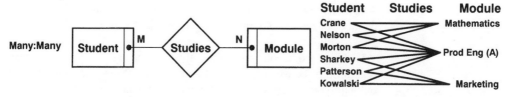

7.4.2 Non-obligatory

1:1. In the example shown in Figure 7.8, clubs must have one (and only one) manager. It is possible, however, for managers to be unattached to any club. Note that the 'blob' associated with the entity 'manager' is outside the box.

Figure 7.8 *Non-obligatory 1:1 relationship*

1:1

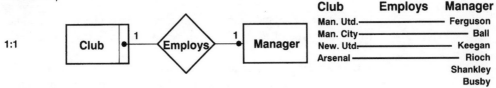

1:Many. In the case of Figure 7.9, postgraduate students (who have no tutor) are included. Similarly, lecturers on sabbatical are included who do not tutor any students.

Figure 7.9 *Non-obligatory 1:many relationship*

1:Many

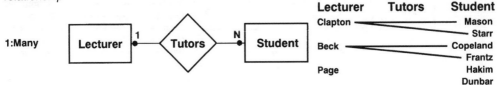

Many:many. In the case of Figure 7.10, the student 'Tracy' is a postgraduate who studies no modules. The module 'Welding Technology' is a final year option that no student has chosen.

Figure 7.10 *Non-obligatory many:many relationship*

Many:Many

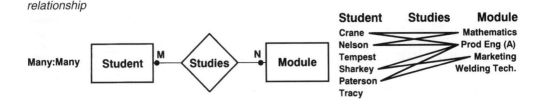

7.4.3 Recursive relationships

A recursive relation is one where a member of an entity is related to a member of the same entity. Examples are as follows:

1:1. In monogamous societies, there is a one to one marriage relationship between people (Figure 7.11). Note that this is not obligatory as 'Geri' and 'Christopher' are unmarried.

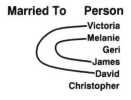

1:1

Person ◁ Married to — Husband (1), Wife (1)

Figure 7.11 *Non-obligatory 1:1 recursive model*

1:many. In this case, note that it is possible for an employee to supervise many people (Figure 7.12). It is impossible, however, for an employee to have more than one supervisor.

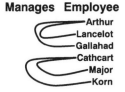

1:Many

Employee ◁ Manages — Subordinate (N), Supervisor (1)

Figure 7.12 *Non-obligatory 1:many recursive model*

Many:many. The recursive many:many relationship is commonly found in Bill of Material (BOM) systems. Figure 7.13, shows two simple BOMs with a total of six parts.

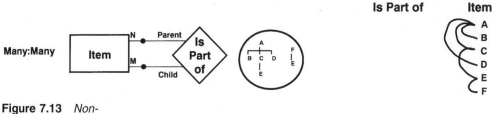

Many:Many

Item ◁ Is Part of — Parent (N), Child (M)

Figure 7.13 *Non-obligatory many:many recursive model*

7.5 Complex models

7.5.1 Graphical conventions

Many information systems within manufacturing industry are highly complex and consist of large numbers of related entities. A simple example is shown in Figure 7.14 for a planning and control system.

7.5.2 Logical errors

For a model to represent an effective system, the logic must be precisely defined. There are two common errors that are often made when defining a model:

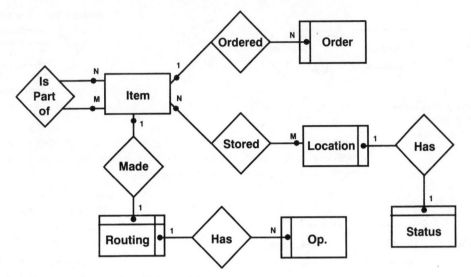

Figure 7.14   *Example of an E-R model*

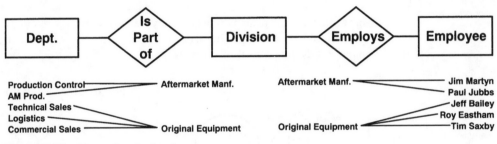

Figure 7.15   *Example of a fan trap*

Fan traps. It might be assumed that Figure 7.15 fully defines the relationships between all of the data elements. In fact, it is impossible to establish which employee works for which department.

Chasm traps. The model in Figure 7.16 avoids the fan trap as described above. If, however, a division directly employs an individual, then the model is inadequate. This problem could be solved by creating a direct

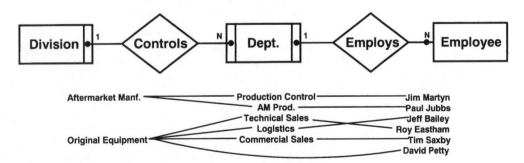

Figure 7.16   *Example of a chasm trap*

link between employee and division, though there are problems with this approach (data redundancy and the potential for inconsistency). Another solution would be to create a 'dummy' department.

7.6 Application of E-R models

E-R models are a formal, graphical technique for defining data structures. They are a powerful conceptual tool for database design. E-R models can form the basis for the detailed specification of database tables.

8 Databases

8.1 Overview

Most modern organizations need to control very large volumes of data. There is no doubt that as business has become more complex, data management has become a crucial element in competitiveness. This chapter will provide an overview of database application and concepts.

8.2 Database types and terminology

There are three broad types of database:

Unstructured. This is a 'collection' of information without a well-defined or predetermined structure. A set of personnel records would be a good example of this type. For each individual various information would be stored – name and address, training information, disciplinary record, photograph, etc.

Simple flat file. This is a simple, but structured set of data. The data is made up of a series of *records*, each with an identical structure. Each record will be made up of a number of *fields* containing specific information. At least one field must be a *key* to identify uniquely the record. This type of database emulates a simple card index structure (Figure 8.1).

Integrated database. This is complex set of data that has internal relationships. This will be covered in greater detail later in this book.

8.3 Indexes

This is a file extract that shows the data in a different order than the original. Design of indexes is critical in manual systems and larger computer databases. In Figure 8.2, two indexes are shown for one file. These indexes allow a database of references to be accessed either by author or by the journal. Note that the physical order of the file is irrelevant if an index is used.

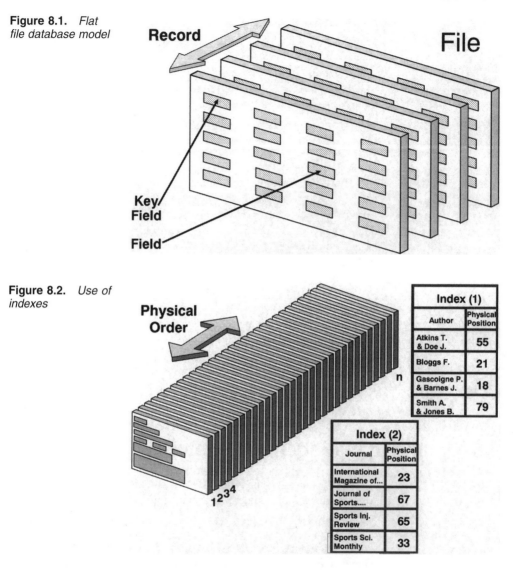

Figure 8.1. *Flat file database model*

Figure 8.2. *Use of indexes*

In complex manual systems (e.g. libraries), indexes are the only means by which data can be accessed in anything other than the physical order.

With computer systems, it is of course possible to interrogate all records. Here, indexes can improve access times very significantly.

8.4 Basic database development

When implementing a database, it is necessary to follow these steps:

- **Fields.** This defines the structure of database records.
- **Layout.** This defines how data will be entered.
- **Output.** These are the reports that will be generated by the systems.

When designing the database, however, the above steps should be completed in the reverse order. This will ensure that the database will meet user requirements. For example, one common mistake even in simple databases is that data are entered inconsistently. For example, the two following codes are different, GOS12457/5 and GOS12457\5, though the reason for this may not be immediately obvious.

Another problem is when particular records are classified by using one or more fields. For example, in a database of vehicles, one individual may describe a particular record as a 'lorry' whereas another might use the term 'truck'. This may not appear a serious problem: it will, however, frustrate any attempt to extract all lorries from the database. Most databases have facilities for using look-up tables on certain fields so that users choose classifications from a menu, rather than directly keying the information. This ensures data are entered in a structured manner, albeit at the cost of additional administration. Another method of improving data accuracy is by correctly choosing field types. Most systems support the following data types:

- **Alphanumeric**. A free format mix of characters.
- **Integer**. Whole numbers.
- **Real number**. A decimal number. In many systems, the number of decimal places to be specified.
- **Date**. This is a calendar date.
- **Logical**. This is a field with only two states: 1 or 0.

8.5 Database management systems

A database management system (DMS) is a piece of software designed to assist in the development of applications. A DMS has a number of key characteristics:

Data structuring. The DMS must allow the relationships between data elements to be structured. This data structure is sometimes referred to as a schema.

Validation and recovery. Since data is independent of the application, this has advantages of consistency and elimination of redundancy. There is the disadvantage, however, that erroneous data entered into the database will affect all aspects of operation. The DMS must have facilities for data validation and recovery following error or corruption.

Data manipulation language. The system must provide a means of accessing the databases and manipulating and updating information.

Most commercially available software for the operational control of manufacturing organizations is based on a DMS. When a DMS is used, data can be considered at three levels of abstraction:

View. This is the view as seen by system users.

Conceptual. This is the level at which the system designer structures the data.

Physical. This is the way data are stored on the computer. This does not have to be considered as part of the system design process, except for providing sufficient storage space.

System designers are concerned with the conceptual level. Data are structured as a series of *tables*.

8.6 Database tables

An example of a database table is shown in Figure 8.3. Notice that each column is headed by an attribute type. Each value within the table is called attribute occurrence. The table is described by the following notation:

Directory (*surname, first name, tel number*)

Note that surname and first name are underlined. This indicates that these fields are identifiers. That is to say, taken together they are unique and fully identify the table row. Earlier in this chapter, these have been referred to as a key. A database table has the following characteristics:

- Order of rows is irrelevant. The order of data as seen by users is a logical view.
- Order of columns is irrelevant. Again, how data are presented to users is a logical view.
- No duplicate entries (keys). No row may be duplicated. This can be ensured by use of a key.
- A row/column intersection has a single value. A 'cell' can hold only a single value.

The table can be said to be well normalized as all attribute values in a row are directly connected.

Normalised Table

Table Name   Attributes   Attribute Type   Attribute Occurrence

Directory	surname	first name	tel number
	Abbott	Louis	0161-745-5000
	Wright	Orville	0161-652-6521
	Cannon	Thomas	0151-362-8947
	Dankworth	John	0121-498-3698
	Everly	Donald	0181-357-9518
	Simon	Paul	0181-357-9518
	Crick	Francis	0171-888-7771

Directory(surname, first name, tel number)

Figure 8.3   *Table nomenclature*

8.7 Normalization and determinants

Normalization is the process of producing efficient database tables. The 'table' shown in Figure 8.4 is invalid because cells have multiple values.

Figure 8.4 *Poorly normalized data table*

Family Group

name	individual	business	domicile	title
Windsor	Philip	Royal Family	United Kingdom	HRH
	Elizabeth			HM
	Charles			HRH
	Anne			HRH
	Andrew			HRH
	Edward			HRH
Corleone	Vito	Organized Crime	United States	Don
	Santino			Mr
	Alfredo			Mr
	Micheal			Don
Tracy	Jeff	Rescue	South Pacific	Mr
	Scott			Mr
	Virgil			Mr
	John			Mr
	Gordon			Dr
	Alan			Mr

Family Group (<u>name</u>, individual, business, domicile, title)

This could easily be rectified by simply duplicating the appropriate family name into the appropriate positions in the table. In this case, the table would consist of a number of repeating groups. This approach, however, has two practical limitations:

Person

name	individual	title
Windsor	Philip	HRH
Windsor	Elizabeth	HM
Windsor	Charles	HRH
Windsor	Anne	HRH
Windsor	Andrew	HRH
Windsor	Edward	HRH
Corleone	Vito	Don
Corleone	Santino	Mr
Corleone	Alfredo	Mr
Corleone	Micheal	Don
Tracy	Jeff	Mr
Tracy	Scott	Mr
Tracy	Virgil	Mr
Tracy	John	Mr
Tracy	Gordon	Dr
Tracy	Alan	Mr

Person (<u>name</u> , <u>individual</u> , title)

Family

name	business	domicile
Windsor	Royal Family	United Kingdom
Corleone	Organised Crime	United States
Tracy	Rescue	South Pacific

Family (<u>name</u>, business, domicile)

Figure 8.5 *Well-normalized tables*

- The table is large and contains duplicated data.
- Data maintenance is cumbersome. For example, if the Tracy family moves to the North Pacific, six rows will need to be updated.

One way to understand this problem is to consider the determinacy relationships within the data. For example, 'domicile' is determined by name as is 'business'. 'title', however, is determined by the composite identifier 'name' and 'individual'. In a well-normalized table (see Figure 8.5, for example) a row does not contain a determinant that is not an identifier.

This is the so-called *Boyce–Codd criterion*. By creating two tables as shown in Figure 8.5, a well-normalized and efficient structure is obtained. The relationship between these two data tables is illustrated in the E-R model in Figure 8.6.

It is also possible to represent the relationship of the two tables in a *determinacy diagram* (Figure 8.7).

Figure 8.6 *Person-family E-R model*

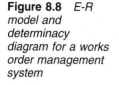

Figure 8.7 *Determinacy diagram for the two tables*

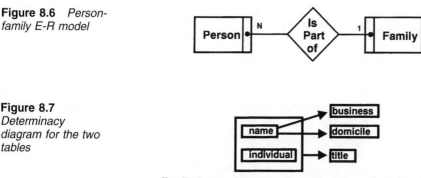

Family (name, business, domicile, Person (individual, title))

8.8 E-R models and databases

E-R models can form the basis for the conceptual design of database structures. Figure 8.8 shows an E-R model and determinacy diagram for a simple works order management system.

Figure 8.8 *E-R model and determinacy diagram for a works order management system*

This is an abstract model; it can be used, however, to structure tables within a DMS (see Figure 8.9).

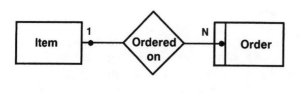

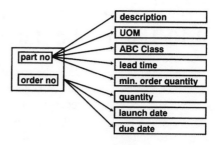

Part Master

part no	description	UOM	ABC Class	lead time	min. order quantity
DP1071	Jaguar Front DBP	EA	A	12	10,000
LN4569	Ford Fiesta Rear MDBL	EA	A	14	20,000
DP2364	Ford Mondeo Front DBP	EA	A	14	10,000
MX1241	High Friction DBP Mix	kg	B	16	635.5

Part Master (part no, description, UOM, ABC class, lead time, min. order quantity)

Order Master

order no	part no	quantity	launch date	due date
123456	DP1071	10,000	01/03/95	13/03/95
123457	LN4569	20,000	01/03/95	15/03/95
123458	LN4569	20,000	08/03/95	22/03/95
123459	LN4569	20,000	15/03/95	01/04/95
123460	DP2364	10,000	08/03/95	22/03/95
123461	MX1241	1271	08/03/95	24/03/95

Order Master (order no, part#, quantity, launch date, due date)

Where: EA = Each   DBP = Disk Brake Pad

MDBL = Moulded Drum Brake Lining   UOM = Unit of Measure

Note: Information on ABC Classes is given in 16.8.1.

Figure 8.9   *Examples of data tables for a works order management system*

8.9 Data table linking

The mechanism of linking tables in the DMS is by means of a common field within the tables (see Figure 8.8). In Figure 8.7, this is accomplished using the field part#. In the case of DMSs, such as Microsoft Access, functionality is available to link data tables together using this approach.

As soon as tables are linked it is essential that users maintain data only via the maintenance framework (screens) created by the system designer. Failure to do this can lead to the database becoming corrupt (though in theory this should be prevented by the DMS). In Figure 8.9, for example, if a part is deleted without checking if a corresponding order exists, then ambiguity will result.

Figure 8.10
Linking data tables

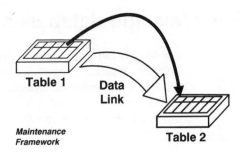

Table 1   **Data Link**

Maintenance Framework   **Table 2**

8.10 Use of the database

Once the structure has been defined within the DMS then suitable data entry/maintenance screens will need to be produced. It will also be necessary to define reports to allow data to be extracted. In practical systems, users will be denied access to the database tables themselves.

8.11 Database summary

The details of how relational databases are created will vary from package to package. There is, however, a set of common principles that can guide the system designer when producing database structures. If these principles are observed rigorously, it is possible to create databases in a formalized manner that are robust and conform to abstract models (such as E–R diagrams).

8.12 System changeover

The final stage in any information system project, whether using bespoke or packaged software, is changeover. This is the transition between the old system and the new. This is a crucial element in the whole process. Even the best-designed system will fail if implemented incorrectly.

There are essentially four methods of system changeover. These are as follows:

Single stage. Here, the new system is implemented in a single step. This approach has the advantage of minimizing effort. There is an element of risk with this approach if a recovery path is not planned.

Parallel. In this case, the old and new systems are run in parallel. This is often considered to be the 'safe' option. In practice, when two systems are run in parallel, the new system never undergoes a thorough test. The main problem with this approach is that the increased workload may be unsustainable.

Phased. Many systems are modular in nature and thus it is possible to phase in the new system. This can create an increased workload in the short term.

Pilot. In this case, the new system in its entirety is implemented in one part of the organization. This is a very effective method as the new system is thoroughly tested. This method can only be applied if a relatively self-contained part of the business can be identified.

8.13 Changeover tasks

The following tasks are generally considered to be part of the implementation process:

Training. This involves the people using the system and possibly external organizations. For example, it may be necessary to inform suppliers that the format of purchase orders will change.

Installation of hardware. This involves the central system resources and also the terminals and printers required by users.

Loading and configuration of software. With complex software, loading and configuration may require considerable effort.

Data entry. This is the entry of data into the system to support its operation initially (see next section).

User support. This is the initial support for all system users during familiarization and identification and correction of teething problems.

The order in which these tasks need to be undertaken will depend on the nature of the implementation.

8.14 Data entry

One of the key stages in system changeover is data entry. In most projects, this is the activity that consumes the greatest effort, at least in terms of sheer man-hours. It should be recognized that there are two data types:

Static data. These are the data that change only slowly (customer addresses, for example). These data can be entered over an extended period prior to implementation. In some cases, this is essential due to the sheer volume of data to be entered.

Dynamic data. These are data that change routinely (customer orders, for example). These changes are sometimes referred to as transactions. These data must be entered immediately before implementation, generally over a weekend.

It is recommended that prior to entering data into a new system, an investigation is made into its accuracy. One of the key issues to be decided is whether all of the data will be entered. Often, it is expedient to leave historical data in paper form if access is required only infrequently. In some cases, it is possible to enter data during operation. For example, it may be decided to enter only customer addresses when an actual order is received.

8.15 Post-implementation tasks

Once systems have been implemented, it is necessary to support them intensively in the short term. With complex systems, teething troubles are common. This may be due to the systems themselves or because of the relative unfamiliarity of users.

Once systems are established, it is common to conduct a post-implementation review. The purpose of this review is to identify any problems in the new systems and produce appropriate solutions.

9 Business process re-engineering

9.1 Overview

One of the most influential ideas in recent years is the concept of business process re-engineering (BPR). This concept was coined by Hammer and Champy (1993) and has been widely embraced by business, particularly in the USA. Indeed, it has been reported that in 1996, 70% of large American corporations were applying or considering applying BPR.

9.2 Nature of change

Organizations are always in a state of change and to cope with this, systems and business practices need to develop. There are, however, two distinct ways in which development takes place as shown in Figure 9.1:

Continuous improvement (CI). Development can take place incrementally, by a process of refinement. This approach is embedded into the total quality management (TQM) philosophy. In Japan, this is

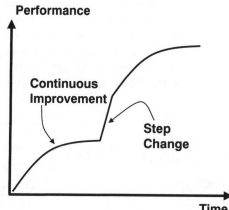

Figure 9.1 *Nature of change*

known as *kaizen* and is most closely associated with Masaaki Imai (1986). The great virtue of CI is that it works because each incremental step is small, hence the chance of radical failure is very small. The disadvantage is that this mode of improvement never makes radical change. Moreover, a 'law of diminishing returns' usually applies to CI; each incremental step has less effect.

Step change. In this case, totally new approaches to problems are employed. By its nature, this involves risk, though the benefits can be extremely large.

At different times, both approaches have a role to play. BPR represents a radical, step change approach and it is supported by a well-developed management philosophy. In one respect, however, TQM and BPR are similar; both have an extremely strong customer focus.

9.3 BPR concepts

9.3.1 Business processes vs functions

'Business process' is an all-embracing term and can be illustrated by Figure 9.2. Business processes can be considered as the interface between the input from a supplier and the output to a customer.

This is easily understood in a manufacturing process. Material is provided by a supplier (input), it is transformed and delivered to a customer (output). Note that the process can be broken down into a number of simpler entities. For example, in order to deliver a product, a design process may be involved.

The model is also applicable to other processes, however. For example, a customer may visit a restaurant. In this case, the output is a combination of a physical entity (the food) and a service. The inputs to the restaurant are provided by a variety of suppliers including vendors of produce and the employees. In this model of a process, it is perfectly natural for customers and suppliers to be *within* the organization itself (so-called 'internal customers' and 'internal suppliers').

Within BPR, a crucial distinction is drawn between processes and functions. Most organizations have traditionally been divided into functional departments. A manufacturing company, for example, would be separated into sales, production, design, accounts, purchasing, etc. Within BPR, functions are far less important than processes. Indeed, the

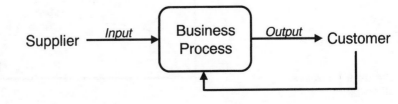

Figure 9.2 *Model of a business process*

Supplier —*Input*→ Business Process —*Output*→ Customer

outcome of a BPR project might be to dispense with the traditional functional management structure.

9.3.2 The BPR approach

BPR, by its nature, is project oriented. That is to say, it has a well-defined start and finish (this is in contrast to *kaizen*). It is undertaken by a project team, usually drawn from a number of functions. BPR can be represented as shown in Figure 9.3.

Figure 9.3 *The BPR approach*

Scope → Import Ideas → Re-Design Process → Plan Transition → Implement

Scope. The first stage is to determine the boundaries of the project. One common danger for *any* project is for objectives to become blurred. Project team can easily stray from addressing one focused issue to attempting to rebuild the entire organization and achieving nothing. One technique that is often helpful is not only to document what is included in the project, but also what is not.

Import ideas. Unlike TQM, the basis for a BPR project is not the existing process (that is not to say, however, that a thorough understanding of existing practices is not essential). It is important to adopt ideas from outside the organization and possibly the industry. For example, many fast-food restaurants have adopted JIT methods for the control of their kitchens.

Redesign process. The next stage is to redesign the process. In the simplest case, this may simply discontinue the existing processes where they have been shown to be redundant. More often, this will involve reorganization and the possible application of IT (IT is seen as a key enabler for BPR).

Plan transition. The nature of BPR projects means that a radical change in operating practices is common. It is necessary, therefore, to carefully consider the implications of the new system to ensure an orderly changeover.

Implement. The final stage is to implement the new project.

The concept of step change is not new and certainly predates BPR. BPR is a complete philosophy, however, and is more than simply a description of the step change process. It contains a number of elements that aim to allow changes to work. The key points of the BPR philosophy can be grouped into three parts: redesign, retooling IT and reorchestration. These are summarized in Table 9.1.

Table 9.1   Three 'Rs' of BPR

Redesign *These are the enabling concepts for process design*	Challenging Eliminating Flattening Simplifying Standardizing Paralleling Empowering Informating Monitoring Partnering Outsourcing Prescheduling	No existing practices are sacred No redundant practices No unnecessary management levels Simple customer/supplier relationships Uniformity in operations Convert sequential processes to parallel Allow employees to make decisions Share information across functions Track key organizational metrics Form strategic alliances with customer and suppliers Focus on core business – eliminate peripheral activities Provide information when known, not when required
Retool IT *IT is a major BPR enabler*	Do current IT systems support business processes? Are new, more appropriate technologies available?	
Reorchestrate *Organizational changes are important to support process changes*	Commitment Incentives Communication Confidence Accountability Obstacles Celebrate	There must be commitment to change at all levels New methods of reward must be devised All staff must understand BPR concepts and reasons for change New systems that do not work will be changed Staff accountabilities must change to reflect changes Organizational practices must not be allowed to hamper change Success should be celebrated to maintain enthusiasm and morale

9.4 Organizational structures for BPR

9.4.1 Hierarchical structures

Classically, organizations (commercial and otherwise) are highly structured. There are well-defined relationships between managers and their subordinates. This is a natural approach with a long history and is well suited to a functional departmental structure. It does, however, suffer from a crucial problem; instructions have to pass down through many levels from top to bottom. This can be time consuming and the clarity of these instructions can be affected. The same problem arises with reporting up the structure.*

This type of structure can lead to individuals being remote from customers (customer focus is very important in the BPR philosophy). A

---

* Further information on organizational structures can be found in *Business Skills for Engineers and Technologists* (ISBN 0 7506 5210 1), part of the IIE Text Book Series.

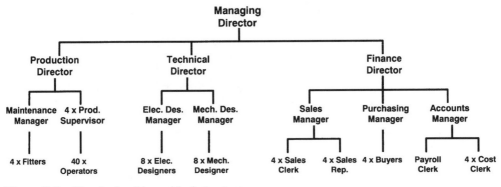

Figure 9.4 *Classical or hierarchical structure*

buyer, for example, may be relatively remote from the customer and focused on his/her own functional department. The rigid nature of the structure can easily destroy initiative. Perhaps most seriously, it can discourage teamwork throughout the organization (and indeed, sometimes encourage departmental hostility). An example of a hierarchical structure is shown in Figure 9.4.

9.4.2 Flattened structure

Many (if not most) companies have reduced the number of levels in the hierarchy. This overcomes at least some of the problems of a highly structured organization: this approach leads to a more responsive organization, for example. It is, however, still oriented towards functional groupings.

9.4.3 Matrix organization

A more radical approach to organization is the application of a matrix structure (sometimes referred to as a network structure). Here, the traditional lines of reporting are dispensed with completely. A common way of doing this is to formulate customer-based teams (see Figure 9.5).

	Production Director		Elec. Des. Manager	Mech. Des. Manager	Sales Director		Accounts Manager	
Customer Manager 1.	Prod. Supervisor	Fitter	2 x Elec. Designers	2 x Mech. Designer	Sales Clerk	Sales Rep.	Buyers	4 x Cost Clerk
Customer Manager 2.	Prod. Supervisor	Fitter	2 x Elec. Designers	2 x Mech. Designer	Sales Clerk	Sales Rep.	Buyers	4 x Cost Clerk
Customer Manager 3.	Prod. Supervisor	Fitter	2 x Elec. Designers	2 x Mech. Designer	Sales Clerk	Sales Rep.	Buyers	4 x Cost Clerk
Customer Manager 4.	Prod. Supervisor	Fitter	2 x Elec. Designers	2 x Mech. Designer	Sales Clerk	Sales Rep.	Buyers	4 x Cost Clerk

Figure 9.5 *Example of a matrix organization*

Under ideal circumstances, this approach has led to significant benefits. In addition, the customer teams in Figure 9.5 may be *self-directed*, so there might be no requirement for a manager. There are, however, potential problems:

Staff can have conflicting priorities. In Figure 9.5, the fitters report to the production director, but have specific responsibilities for customer teams. If a fitter is ill, should the fitters in other teams be responsible for cover?

Problems in existing companies. In Figure 9.5, staff are distributed conveniently; for example, there are four buyers and four customer teams. What would happen, however, if a company moving to this structure had either three or five buyers?

Difficult to manage service functions. If there are service functions, like personnel or management information services (MIS), how will these fit into the matrix structure? One solution advocated by the BPR movement is outsourcing – i.e. allowing this function to be performed by another company.

Reward/promotion structure. Traditionally, people have expected to be rewarded for excellent performance by being promoted within the hierarchical structure. In a matrix structure, new means of rewarding employees need to be devised to retain staff.

9.5 BPR – summary

There are two modes of change: continuous and step. This is not a new idea, but the BPR and TQM approaches provide a comprehensive management philosophy to these change mechanisms. It must be recognized that BPR can be a difficult undertaking and changes to the organization are often required to support new business processes. Many companies have found this challenging. Certainly, to regard BPR as a panacea would be mistaken.

BPR is often thought of as an IT/IS technique. While IT/IS is a key enabler in many projects, BPR is a management rather than a technical approach. Finally, along with TQM, BPR is one of the most important management ideas of recent times.

9.6 Case study

9.6.1 Avoiding overcomplexity

Saiful Pumps Ltd manufactures a range of small electric pumps for industrial applications. The organization employed a large number of people in its commercial and technical sales departments. Commercial sales would initially receive enquiries and the potential customer entered onto the database (if they were not already on file). The customer's requirements were passed to the technical sales department where it was determined whether a standard pump could be used or if engineering work would be required. When this had been done, a report was passed back to the commercial sales office who would advise the customer of the quoted price. Because of the delays that were incurred in passing paperwork between the two departments, it was decided that the company should invest in appropriate computer systems. A project team was formed with representatives from commercial and technical sales as well the company's IT/IS department.

The project team quickly concluded that there were several fundamental problems. First, many of the delays were the result of documents having to be passed between two separate offices. Second, 80% of the order requests were eventually satisfied by a standard pump, with only 20% needing a special pump to be designed. In many of these cases where a standard pump was needed, this was obvious to commercial sales without the need for a design engineer to be consulted (e.g. if a customer specified the pump needed via its catalogue number).

The project team moved the commercial and technical sales teams so they were based in the same open-plan office. Next, they implemented a new procedure for handling orders which was christened 'the Fast-Track Pump System'. When an enquiry was received, a commercial sales clerk would make an initial determination as to whether a technical sales engineer needed to be consulted. If it were clear that a standard pump was required, an immediate reply would be made to the customer (this accounted for almost 50% of enquiries). If the commercial sales clerk was almost certain a standard motor could be used, they could consult a technical sales engineer at a desk nearby. Over time, the commercial sales clerks became more adept at identifying where standard pumps could be used.

9.6.2 Analysis

While the project also introduced computer-based order process-ing systems, the project team believed most of the improvement was the result of the relatively simple measures they had put in place.

Exercises – Information systems in manufacturing

Exercises – Introduction to systems analysis

1 Osbourne Ltd is a manufacturer of simple suspension mount assemblies based in the North West. At the time of the case study, the company had roughly 1200 employees and a turnover of £100 M. Most of the senior management team had been in post for many years, but a new managing director (MD) joined the company at the time of this case study. He quickly identified that the company had a highly centralized organizational structure. This was based on functional departments – production control, sales, purchasing, accounts, etc. The company served four distinct markets however:

- **Original equipment (OE).** The company sold suspension assemblies to automotive original equipment manufacturers (OEMs) such as Ford, Opel and VW-Audi Group (VAG). This market represented approximately 35% of turnover. These products were manufactured in relatively high volumes and low varieties, as the vehicles on which these assemblies were to be used were in current production.

- **Spares distribution.** The company had a warehouse that distributed automotive spares to customers such as Halfords and Motorworld. This market represented approximately 35% of turnover. Around half of the products sold were produced internally (see next point). The remainder were procured from other parts of Osbourne's parent group.

- **Spares manufacture.** The company produced assemblies for aftermarket application. Virtually all of the products manufactured were sent to the company's warehouse (see above). This market represented approximately 20% of turnover. Products were manufactured in relatively small batch sizes as the vehicles on which these assemblies were to be used were *not* in current production. As a result, separate production equipment was required for OE and spares manufacture.

- **Railways.** The company also made products for the niche train/tram market using specialized equipment. This market represented approximately 10% of turnover.

The company had been a pioneer in the use of IT in manufacturing. By the time of the case study, however, the systems were relatively old. All of the software had been written in-house. The systems were based on the parent group's mainframe computer. Usage charges were based on computer processing time. In the year of the case study, mainframe charges amounted to £200 K.

The company had a systems department with ten people. Most of their time was spent on solving problems with the software and writing special reports for the management team. The data processing (DP) manager stated that this was due to the fact that systems had evolved over a long period. Programs were added on top of existing systems and this led to great complexity with the attendant possibility of errors. Very little time was available for the development of new systems. Where new development was undertaken, this invariably meant adding new programs to existing systems. This was because there was insufficient time for fundamental re-design.

Because of the above, many business areas had no computer support and many users had developed their own systems. The most important and sophisticated of these was for quality management and was based on a PC network. This had been developed by one of the quality engineers who had had a heart attack and therefore could not perform his previous shop floor role. Because he had a computer at home, however, he volunteered to write the quality database. This collected important quality data for batches of manufactured products and raw material. Other company functions simply used manual systems. This caused some complaints from users. For example, maintenance engineering used card-based records. As a result, they were unable to run any form of preventive maintenance.

The MD proposed a strategy to his senior management team. First, the company should be organized into four market-focused divisions. Each would be profit-accountable and have a significant amount of autonomy. The centralized functions such as production control would be split up and personnel allocated to the divisions. Thus each division would have its own sales departments for example. The only centralized functions were to be catering, human resources management (HRM) and buildings maintenance. This would represent a radical and dramatic change in the operation of the business. The MD believed this was crucial in order to remain competitive.

To support this strategy, the company would replace the mainframe systems with an integrated business management package running on a mini-computer based on-site (such a package would be described as an enterprise resource planning (ERP) system, which will be described in 17.5). Two vendors of ERP packages made presentations to the senior management team. A number of objections were raised, however:

- The total cost of the ERP package would be in the region of £1 M.
- The ERP package was not tailored to the business in the same way as the existing bespoke systems. The production manager made the following comment, 'The ERP package tracks batches of work like the mainframe system. On the screen, however, it asks for a "works order number" to enter the information. The operators will not understand this; they are used to the phrase "factory batch number". All of the terminology on the ERP system is wrong.' Most of the senior management team agreed with this view. While ERP packages had been successfully employed elsewhere in the parent group, the senior management team felt that Osbourne was unique and thus a standard piece of software was inappropriate.
- Most of the systems department was relatively old. While the DP manager acknowledged fewer people would be needed (around five) to run the ERP package, he felt only two of his younger staff would be willing and able to re-train. The others were nearing retirement age and would therefore lack motivation.

An alternative strategy was proposed. The systems department should recruit five new members of staff. Additional programs would then be written and added onto the existing systems. These would have the objective of allowing divisional operation. Once this was complete, the business could move to a divisionalized structure. The DP manager estimated that this would take three to four years. This approach would be far less dramatic than the alternative. Indeed, this approach would allow a gradual move towards the divisional structure.

After reading the information above, what advice would you give to the MD? Write this advice in the form of a short document (150–300 words).

2 Northway Engineering Ltd manufacture a wide range of valves and fittings for use in the oil industry. The company have an annual turnover of £20 M and employ 300 people. The company have utilized information technology in several application areas for a number of years. At present there are several separate hardware platforms in use:

- **Mini-computer with Unix operating system.** This is connected via a local area network (LAN) to a number of departments. This computer is provided with a range of software to cover the requirements of production planning and control, purchasing, accounting, payroll and personnel.
- **Design engineering network.** This comprises four workstations running the Unix operating system linked with each other using a LAN. This network is provided with software to enable it to be used for 3D modelling, stress analysis, flow analysis and other design calculations and computer-aided draughting (CAD).
- **Production engineering department.** There are two stand-alone PCs, one used for computer-aided NC part programming and one for jig and tool design and production of plant layout drawings.

- **PCs in quality assurance department.** A stand-alone PC is used for analysis of reject/rework figures, finished product reject analysis and warranty claims. A further two PCs are used on the shop floor linked to in-process gauging equipment for statistical process control.
- **PC in maintenance department.** This stand-alone PC is used to run a planned maintenance scheme; it is also used for stock control of spare parts.
- **PCs for clerical and administrative tasks.** There are a considerable number of stand-alone PCs in use throughout the company for word processing and other tasks. At present these computers are not networked.

Some of the above systems are due to have either hardware or software upgrading. Senior management have received a number of separate requests from departmental heads for an allocation of funds to carry out improvements within their own departments. The chief executive of the company is concerned that there appears to be no coordination between departments and has asked if a move to integrating these subsystems has been considered.

Discuss how you would approach the task of assessing the need for the integration of the computer systems used within the company. The role of data exchange should be discussed as well as the distinction between 'interlinking' and 'integration'.

Exercises – Systems design

1 A manufacturing company completes product that is then shipped to customers. The company has a computer system to support this process. The process is as follows:

(a) The production department completes batches of product that are received into finished goods stock (FGS) by the storeman. Each batch has a document called a works order that the storeman uses to update computer-based stock records when product is received.

(b) The sales department maintains customer order (CO) records on the computer. They review stock records and where a CO is due and stock is available, the sales department instructs the storeman to ship product.

(c) The storeman transfers product from FGS to the despatch area to be packed for shipment. The storeman produces a packing note that states how the product is to be packaged. The storeman then updates stock records and CO records.

(d) The despatch department packs the products and writes out a shipping note. The packaged products (with the shipping note attached) are placed on a lorry and sent to the customer.

Assuming the production department, sales department, despatch area and customer are external entities, draw the following for the process described in (a) to (d) above:

(a) A data flow diagram (DFD).

(b) A resource flow diagram (RFD).

Exercises – Data analysis

1 A university engineering school is designing a database. The database needs to control the following entities:
   - PhD students.
   - Undergraduate students.
   - Lecturers.
   - Units (of study).

   Each PhD student is supervised by a lecturer. Some lecturers have as many as four PhD students though some have none. Each unit is led by one lecturer. Some lecturers, however, do not lead any units. Each undergraduate student studies several units. Each unit is studied by a number of undergraduate students. Finally, each undergraduate is tutored by a lecturer. Some lecturers do not tutor any students.

   Draw an entity-relationship (E-R) diagram to represent the situation described above.

2 A small engineering company plans to develop software to automate some elements of its manual design processes.
   (a) Explain why it would be inadvisable to start the process by generating program code.
   (b) If the development team wished to understand the existing systems, discuss the ways in which information could be gathered that would be appropriate in this case.
   (c) Outline the advantages of using a formal method of system representation (such as data flow diagrams) rather than a natural language (such as English).

3 A university is designing a database. The following E-R diagram has been proposed for part of the design:

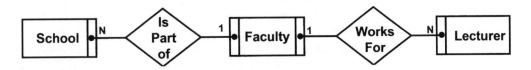

   The university has five faculties and 15 schools. A sample of the data to be entered is shown below:

School	is part of	Faculty
Physics		Science
Chemistry		
Biology		
Engineering		Technology
Computing		
Textiles		

   If a database was developed from the diagram, what problems might arise? Sketch an improved E-R model.

Exercises – Databases

1 A small company holds the following data relating to external contacts in the table shown below:

Name	Initial	Title	Extension	Company name	Switchboard number	Address
Atkinson	J	Buyer	5934	Ariston Engines	0121-294-4856	25 Jacobs Way, Birmingham, B83 4DF
Bailey	J	Sales Clerk	3334	Bearson Pumps	0141-395-4967	106 Millius St, Glasgow, G38 5HH
Crossley	R	Accounts Clerk	214	Fogg International	01142-451455	233 Highfield Lane, Sheffield, S11 9JK
Degg	R	Sales Clerk	1021	Diamond Gears	0161-334-6767	12 Horners Road, Manchester, M33 7GH
Ekere	N	Buyer	214	Fogg International	01142-451455	233 Highfield Lane, Sheffield, S11 9JK
Forsyth	D	Accounts Clerk	1021	Diamond Gears	0161-334-6767	12 Horners Road, Manchester, M33 7GH
Gee	G	Designer	214	Fogg International	01142-451455	233 Highfield Lane, Sheffield, S11 9JK
Hinduja	N	Buyer	4457	Ariston Engines	0141-294-4856	25 Jacobs Way, Birmingham, B83 4DF
Ingle	C	Designer	1021	Diamond Gears	0161-334-6767	12 Horners Road, Manchester, M33 7GH
Johnson	C	Designer	4457	Ariston Engines	0141-294-4856	25 Jacobs Way, Birmingham, B83 4DF
Kik	B	Sales Clerk	1021	Diamond Gears	0161-334-6767	12 Horners Road, Manchester, M33 7GH
Labib	A	Accounts Clerk	3334	Bearson Pumps	0414-395-4967	106 Millius St, Glasgow, G38 5HH
Mohamed	S	Designer	4457	Ariston Engines	0141-294-4856	25 Jacobs Way, Birmingham, B83 4DF

Normalize the data above and thus create two new tables.

2 In relation to DMS briefly discuss:
   (a) Data structuring.
   (b) Validation and recovery.
   (c) Data manipulation language.

3 Discuss the changeover and data entry strategies that could be applied to the following cases:
   (a) A small company is implementing a simple 2D computer aided design (CAD) system. Previously, all draughting was manual. Roughly ten new drawings are created per week. The company has over 10 000 paper-based drawings.

(b) A large healthcare company has 25 hospitals. The current systems for storing patient records is based on a central mainframe computer. These patient records are amended frequently and thus transaction rates are very high. It is critical the system is available at all times. The company plans to decentralize systems with each hospital having its own independent mini-computer based systems.

4 A small independent bookshop is computerizing its catalogue using a simple flat file database on a PC.

(a) Explain why it is necessary to define a key field. What would be a suitable choice in this case?

(b) One of the fields required on the database relates to the type of book (e.g., 'travel', 'sport', 'science fiction', etc.). Why would it be useful to employ a look-up table on this field?

(c) The shop owner would like to store the chapter names for each of the books in the catalogue. Explain why this would be impossible with a flat file database.

Exercises – Business process re-engineering

1 Give a reasoned argument as to why a company should employ BPR rather than simply computerizing its existing systems.

2 Briefly outline the key areas of similarity and difference between business process re-engineering (BPR) and continuous improvement within total quality management (TQM).

3 Compare and contrast BPR and soft systems methodology.

Part 3

Quantitative Methods

Case study

Decisions, decisions

Tricia and Melanie were sisters. They were both studying engineering at university in the same city and were sharing a flat. They were about to go out to a party, but couldn't agree upon how to get there. They'd both agreed that they needed to save money that month and were trying to work out the cheapest way of getting to the party.

'Well, I think we should take the car', said Tricia.

'But we can get a lift to get there and there's a good chance we'll get a lift back. It'll cost at least £2 if we take the car', said Melanie.

'£2 – big deal. What if we can't get a lift back? Then we'll have to call a taxi and that'll cost £7.50.'

'Well, I think there's a four in ten chance of getting a lift back, it's got to be worth taking the chance. If we can get a lift back it won't cost anything.'

'Where does "four in ten" come from?'

'It's based on the last 50 parties we've been to this year.'

'Well if it's four in ten, then there's a six in ten chance we won't get a lift. Then we'll have to pay £7.50.'

'OK, OK. We'll take the car then – let's just go.'

'I'll drive there, you drive back.'

'No. You drive back. I was the designated driver last time.'

'How about heads or tails?'

Analysis

There is a rational way of approaching this problem. The solution lies in balancing the benefits of a particular outcome occurring. A detailed solution to the problem is presented in 13.3.

Summary

This part reviews a range of quantitative techniques that can be used by manufacturing managers. The part starts with a review of some basic concepts in probability and statistics. Forecasting techniques are then reviewed. Forecasting is essential for all organizations and this is particularly true of manufacturing businesses. Next, optimization techniques are described with particular emphasis on linear programming.

All managers need to make decisions (indeed, decision making characterizes the management process). This part will describe techniques that can be used to support the decision-making process. The part also discusses simulation which is an increasingly common technique in manufacturing industry. Finally, this part will outline project-planning methods.

Objectives

By the end of this part, the reader should:

- Appreciate the importance of quantitative techniques.
- Understand the main concepts of the techniques described.
- Be able to apply the techniques to solve problems that might be typically encountered in a manufacturing organization.

10 Probability and statistics

10.1 Overview

10.1.1 Qualitative and quantitative analysis

Analysis is defined as the breaking down of systems into their constituent parts. There are two ways in which this can be accomplished:

Qualitative analysis. This is characterized by the description of a system in non-numerical terms. A simple case of qualitative analysis would be the description of an individual as tall or short. Qualitative analysis, therefore, often involves judgement on the part of the analyst.

Quantitative analysis. This is characterized by describing a system in numerical terms. For example, a person's height might be measured as 1.80 m. Note that this is an objective measurement, rather than a subjective judgement. In some texts, quantitative analysis is referred to as operations research or management science.

One of the critical skills required by manufacturing managers is the ability to make decisions. This part of the book will review some of the tools that can be applied to assist in the decision-making process. It should be noted, however, that while quantitative analysis techniques can be extremely valuable, the results will always need to be tempered by the experience and judgement of the manager concerned. The quantitative analysis can be represented as shown in Figure 10.1.

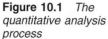

Figure 10.1 *The quantitative analysis process*

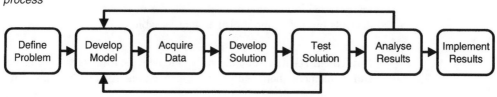

Define problem. By the nature of quantitative analysis, it is essential to formulate a clear and unambiguous statement of the problem. In many respects, this is the most important and difficult step in the entire process. A number of authors (in particular, Checkland with his SSM as discussed in 6.11) have suggested that so-called *hard systems* approaches (such as quantitative analysis) are fundamentally flawed as the nature of the solution is fixed at too early a stage.

Develop model. Models can take several forms (e.g. physical or schematic models are commonly used in engineering). This chapter, however, will focus principally on logical or mathematical models.

Acquire data. It is crucial when conducting quantitative analysis that accurate input data are obtained. This can be extremely difficult in some cases, particularly if the problem is large or complex. The nature of the data collection task can vary from problem to problem. In manufacturing problems, it is likely that interviews with key personnel will need to be conducted.

Develop solution. Once the data have been entered into the model, it is possible to develop an optimal solution. This may be achieved in essentially two ways:

Algebraic solution. Here, the model consists of one or more equations that are solved to provide an answer.

Numerical solution. In this case, a variety of alternatives are tested until a viable answer is produced. In manufacturing, a common example of this case is *discrete event simulation*. It is common for computers to be applied to solve problems numerically.

Test solution. Before implementing the results of analysis, it is important to verify the validity of the model and input data. It is essential to check that the answers given are consistent with the problem overall. One method that can be applied is to try simple or known cases on the model and check the results are reasonable. If the results show that flaws exist, then the model development process needs to be revisited and refinements made.

Analyse results. The next stage is to determine the implications of the results. One common practice is to undertake *sensitivity analysis*. This involves using the model to establish the degree to which the model depends on one or more of the input variables.

Implement results. The final stage is to take the results from the model and implement them within the organization. Again, this can prove difficult in practice.

10.2 Probability

10.2.1 Basic definitions and theory

Before exploring quantitative methods for manufacturing, it is worthwhile reviewing basic theory in probability and statistics. To some extent, this part of the book discusses ways in which an uncertain world can be analysed. A logical starting point therefore is an examination of probability theory.

There are two basic rules of probability. First, all probabilities lie in the range 0–1. If the probability of an event is $P(e)_i$, then:

$$0 \leq P(e)_i \leq 1 \tag{10.1}$$

Second, the sum of all possible outcomes is equal to 1. Expressing this algebraically:

$$\sum_{i=1}^{i=n} P(e)_i = 1 \tag{10.2}$$

Crucial to probability theory are the concepts of *trials* and *events*. An example of a trial is drawing a card from a shuffled 52-card deck. There are a number of events that can be tested, however. For example, the probability of drawing a diamond is 1/4 (0.25) and a seven 1/13 (0.077). When considering events, two concepts are important:

- **Mutually exclusive events.** In this case, only one of the events can occur in any trial. For example, drawing a club and diamond is mutually exclusive: a card cannot be both a club and a diamond. On the other hand, drawing a seven and a diamond is not: the card could be the seven of diamonds.
- **Collectively exhaustive events.** Here, the events represent all of the possible outcomes. For example, drawing a face and a number card is collectively exhaustive: a card must be one of the two alternatives. Drawing a black card and a spade is not collectively exhaustive as in the case of drawing the three of clubs.

10.2.2 Addition of probabilities

If events are mutually exclusive, then adding probabilities is simple. If there are two events, a and b, then the probability of either occurring is given as:

$$P(e_a \text{ or } e_b) = P(e_a) + P(e_b) \tag{10.3}$$

Thus the probability of drawing either a diamond or a heart is:

$$P(\blacklozenge \text{ or } \blacktriangledown) = P(\blacklozenge) + P(\blacktriangledown)$$

$$= \frac{13}{52} + \frac{13}{52} = \frac{1}{2} = 0.5$$

There is a complication, however, if events are not mutually exclusive, 'double counting' occurs. In this case, the probability of two events is given by:

$$P(e_a \text{ or } e_b) = P(e_a) + P(e_b) - P(e_a \text{ and } e_b) \quad (10.4)$$

Consider the case of drawing a diamond or a five. At first glance, the probability of either of these two occurring is $1/4 + 1/13$. This fails to take into account the fact that one card is included in both cases (the five of diamonds) and is counted twice. Thus the correct expression is as follows:

$$P(\blacklozenge \text{ or } 5) = P(\blacklozenge) + P(5) - P(\blacklozenge \text{ and } 5)$$

$$= \frac{13}{52} + \frac{4}{52} - \frac{1}{52} = \frac{16}{52} = 0.308$$

10.2.3 Independent events

It can be seen from the previous discussion that the number of possible ways a particular event can occur divided by the total number of possible outcomes gives the probability of any event occurring. Thus if a coin is tossed, there is only one way of getting a head. The total number of possibilities, however, is two (heads or tails). Thus the probability of getting a head is $1/2$. This is described as *marginal* or *simple probability*.

In a series of trials of tossing a coin, the outcomes are independent. That is to say, one throw does not affect any other. This is often misunderstood. The so-called *gambler's fallacy* is that if a large number of heads occur, then the next throw is more likely to be a tail to 'even out the luck'. This is not true: over time, the number of heads and tails will even out, but only because of the increasing number of trials. It is not because of any influence between one event and another.

Consider now the joint probability of two independent events a and b as in tossing two coins and getting two heads. By intuition, the probability is $1/2 \times 1/2 = 1/4$. Mathematically:

$$P(e_a e_b) = P(e_a) \times P(e_b) \quad (10.5)$$

10.2.4 Statistically dependent events

Some events are *not* independent. According to proverb, the probabilities of a red sky occurring at night and fine weather the next day are not independent. This gives rise to the notion of *conditional probability*: i.e. the probability of an event occurring once some other event has already taken place. If events are independent, such as the toss of a coin, conditional probability can be expressed simply:

$$P(e_a \mid e_b) = P(e_a) \tag{10.6}$$

$$P(e_b \mid e_a) = P(e_b) \tag{10.7}$$

Note that the '|' symbol means 'conditional on'. Thus if a head is thrown, there is a 0.5 probability subsequently of throwing a tail and vice versa. If the two events are not independent, the situation is more complex. Consider the case below:

	Spot	No spot
Blue balls	10	20
Red balls	6	24

There are 60 balls in a bag, 30 blue and 30 red. Of the blue balls, ten have a spot. Of the red balls, six have a spot. Looking at the individual probabilities:

$$P(e_{Blue}) = \frac{30}{60} = 0.5 \qquad P(e_{Spot}) = \frac{16}{60} = 0.267$$

$$P(e_{Red}) = \frac{30}{60} = 0.5 \qquad P(e_{Plain}) = \frac{44}{60} = 0.733$$

If a red ball is drawn, what is the probability that it will have a spot? *Bayes' theorem* can answer this question. This can be expressed mathematically as follows:

$$P(e_a \mid e_b) = \frac{P(e_a e_b)}{P(e_b)} \tag{10.8}$$

Expressing the question mathematically:

$$P(e_{Spot} \mid e_{Red}) = \frac{P(e_{Spot} e_{Red})}{P(e_{red})}$$

Thus:

$$P(e_{Spot} e_{Red}) = \frac{6}{60} = 0.1 \qquad P(e_{Spot} \mid e_{Red}) = \frac{0.1}{0.5} = 0.25$$

10.2.5 Probability trees

So far, only a maximum of two events has been discussed. In more complex situations it is useful to use probability trees. Consider four coins being tossed in succession. As shown in Figure 10.2 in this case there are 16 possible outcomes:

Note that at each branching point, the probabilities add up to one in accordance with equation (10.2).

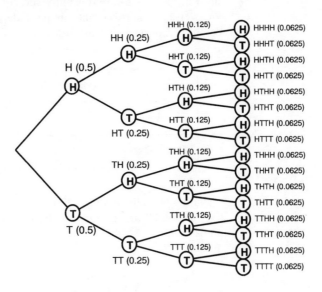

Figure 10.2
Probability tree for four coins

Key: H = Head   T = Tail

10.3 Probability distributions

10.3.1 General theory

In Figure 10.2, if a head is worth 1 and a tail 0, then there is only one way to score 4 (HHHH) hence a probability of 1/16. Similarly, there is only one way to score 0 (TTTT). There are six ways, however, to score 2 (HHTT, HTHT, THTH, THHT, HTTH and TTHH) hence a probability of 6/16. If the score is plotted against probability, a probability distribution is produced as shown in Figure 10.3.

The concept of a probability distribution is extremely valuable. It allows the probability of a particular outcome to be determined without the need to generate the probability tree.

10.3.2 The normal distribution

There are a number of standard probability distributions. The one depicted in Figure 10.3 is a *binomial distribution*. The commonest type

Figure 10.3
Probability distribution

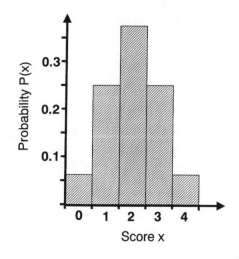

of distribution is the *Normal or Gaussian distribution*. Unlike the binomial distribution, it can be applied to continuous rather than discrete variables. The general form of the distribution is shown in Figure 10.4.

The shape of the curve is given by the formula (the derivation of this expression is beyond the scope of this chapter):

$$f(x)^2 = \frac{1}{\sigma\sqrt{2\pi}}\, e^{\frac{-\frac{1}{2}(x-M)^2}{\sigma^2}} \tag{10.9}$$

where: σ = standard deviation, x = variable and M = mean.

The area under the curve as shown in Figure 10.4 gives the probability of a particular outcome lying between two values x_1 and x_2. Mathematically:

$$P(x_1, x_2) = \int_{x_1}^{x_2} \frac{1}{\sigma\sqrt{2\pi}}\, e^{\frac{-\frac{1}{2}(x-M)^2}{\sigma^2}}\, dx \tag{10.10}$$

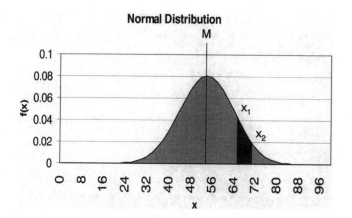

Figure 10.4 *A normal distribution*

Note that for a normal distribution:

- $+/-\ 1\sigma = 68\%$ of the possible outcomes.
- $+/-\ 2\sigma = 95.4\%$ of the possible outcomes.
- $+/-\ 3\sigma = 99.7\%$ of the possible outcomes.

10.3.3 Standard statistical formula

Referring to the Normal distribution and the equations above, the following formula can be used. These will be applied in the rest of the book:

$$\text{Mean } (M) = \frac{1}{n} \sum_{i=1}^{i=n} x_i \tag{10.11}$$

$$\text{Standard deviation } (\sigma) = \sqrt{\frac{1}{n} \sum_{i=1}^{i=n} (x_i - M)^2} \tag{10.12}$$

$$\text{Mean absolute deviation (MAD)} = \frac{1}{n} \sum_{i=1}^{i=n} |(x_i - M)| \tag{10.13}$$

This formula is modified when applied to a forecast (see Chapter 11).

$$\text{MAD} = \frac{1}{n} \sum_{i=1}^{i=n} |(x_i - F_i)| \tag{10.14}$$

$$\sigma = \sqrt{\frac{\pi}{2}} \text{ MAD} \quad \text{Normal distribution only} \tag{10.15}$$

where n = number of data elements, x = a particular data element and F_i = forecast.

11   Forecasting

11.1 Overview

Forecasting is required at a number of levels in manufacturing organizations. Essentially, forecasting has two purposes. First, to provide general information to support decision making. Second, to anticipate changes so that appropriate action can be taken. Forecasting is required over a range of time periods.

- **Short term.** Short-term forecasts are typically highly detailed. For example, a retail organization will need to forecast sales for each stocked item to facilitate effective replenishment. In many cases, short-term forecasts will be generated automatically (for example, by *exponential smoothing*).
- **Medium term.** Medium-term forecasts are less detailed. Taking the example of a retailer, it may be necessary to forecast future consumer trends so new items can be introduced. These forecasts are less detailed, but will require greater management attention (for example, to identify new suppliers).
- **Long term.** Long-term forecasts are concerned with important trends in the relatively distant future. For example, a retail organization will need to forecast consumer behaviour over the long term to plan the building of new stores.

There are three basic methods of forecasting: intuition, extrapolation and prediction. These will be discussed in turn.

11.2 Intuitive forecasting

Intuitive forecasts are a subjective assessment of future events. There are three ways in which this type of forecasting can be undertaken:

- **Individual.** This is a forecast undertaken by a single individual without reference to other sources of information.

- **Conference/Survey.** A group of company employees may partici-
  pate to jointly arrive at a forecast. Alternatively, a survey of
  outsiders (normally customers) may be undertaken.
- **Delphi.** This involves a group independently and anonymously
  forecasting. The results are then shared and a second round of
  forecasting is undertaken.

11.3 Extrapolation

11.3.1 Overview

Extrapolation is an analytical approach to forecasting. The fundamental
premise of this approach is that the future can be forecast by an
examination of the past. Extrapolation is therefore based on previous
events for a *single* variable. The simplest method is graphical: that is to
say past data are plotted on a graph and a forecast generated by
inspection. A simple, analytical method is the *moving average*. For
example, a company might average the last ten weeks' sales to forecast
sales for next week. Mathematically, this can be expressed as
follows:

$$F_{(i+1)} = \frac{1}{n} \sum_{i=(n-k)}^{i=n} X_{(n-i)} \tag{11.1}$$

Where n = number of data points, k = number of points used to average,
x_i = data element and $F_{(i+1)}$ = forecast for next period.

11.3.2 Exponential smoothing

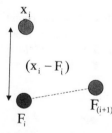

Figure 11.1 *1st-order smoothing*

A more elegant approach is exponential smoothing. This is a weighted
moving average where the most recent data elements are judged to be
more influential. Analytically, referring to Figure 11.1, it can be
expressed as follows:

$$F_{(i+1)} = \alpha x_i + (1 - \alpha) F_i$$

$$F_{(i+1)} = \alpha x_i + F_i - \alpha F_i$$

$$F_{(i+1)} = F_i + \alpha(x_i - F_i) \tag{11.2}$$

where α = smoothing factor.

Smoothing factors are normally set in the range 0.1 to 0.3. The greater
the smoothing factors, the more erratic the forecast. If α is set to unity,
the forecast will simply track the demand, lagging by one period. If α
is set to 0, the forecast will remain at a constant value. The graph in
Figure 11.2 shows how exponential smoothing would track a demand
for a particular sales pattern.

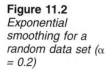

Figure 11.2
*Exponential
smoothing for a
random data set (α
= 0.2)*

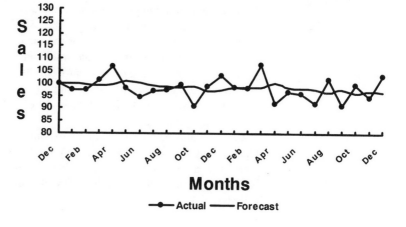

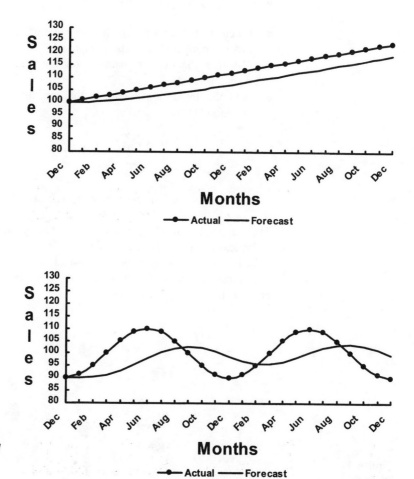

Figure 11.3
*Exponential
smoothing for trend
and seasonal
patterns (α = 0.2)*

Simple smoothing approaches are vulnerable to trend as the forecast lags the demand as shown in Figure 11.3. Similarly, these approaches are also vulnerable to seasonality for the same reason.

The basic premise of exponential smoothing is that the forecast will move toward the actual data element from the previous forecast value. To provide an element of smoothing, however, this movement is attenuated by the factor α. Large values of α will cause the forecast to be erratic and impractical to apply.

As stated above, however, simple exponential smoothing is limited when dealing with data with an in-built trend. Second-order exponential smoothing can be applied more effectively in these cases. This approach examines the data for underlying trend and modifies the forecast accordingly. In effect, second-order exponential smoothing 'anticipates' trend.

11.3.3 Second-order smoothing

The principles of second-order smoothing are shown in Figure 11.4. The following stages need to be followed:

- Calculate the forecast for the next period, $F_{(i+1)}$, using the first-order exponential smoothing approach.
- Calculate the trend adjustment for the next period, $T_{(i+1)}$, using equation (11.3).
- Add the trend adjustment to the forecast, $F_{(i+1)}$, as shown in equation (11.4).

$$T_{(i+1)} = T_i + \beta(F_{(i+1)} - F_i) \tag{11.3}$$

$$F'_{(i+1)} = F_{(i+1)} + T_{(i+1)} \tag{11.4}$$

where β = second-order smoothing factor.

The graph in Figure 11.5 illustrates the use of second-order smoothing.

Typically, β lies between 0.1 and 0.3. Second-order smoothing is less effective when dealing with seasonal trends (see Figure 11.5). Techniques are available for seasonal correction, though these are beyond the scope of this book.

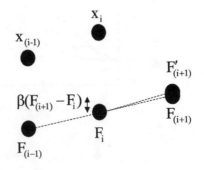

Figure 11.4 *2nd-order smoothing*

Figure 11.5 *2nd-order smoothing for data sets with trend and seasonality α = 0.2, β = 0.2)*

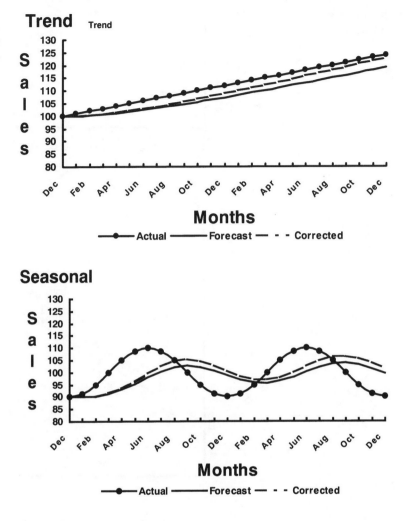

11.4 Prediction

11.4.1 Overview

Prediction is an assessment of the future based on a number of variables. Unlike extrapolation, prediction examines present conditions to forecast the future. For example, there is a relationship between the average cost of houses in the future and present interest rates. Prediction therefore depends on understanding the relationship between different variables.

11.4.2 Regression analysis

Consider the graph in Figure 11.6. This shows the final module marks of a group of students plotted against their attendance at classes. Two

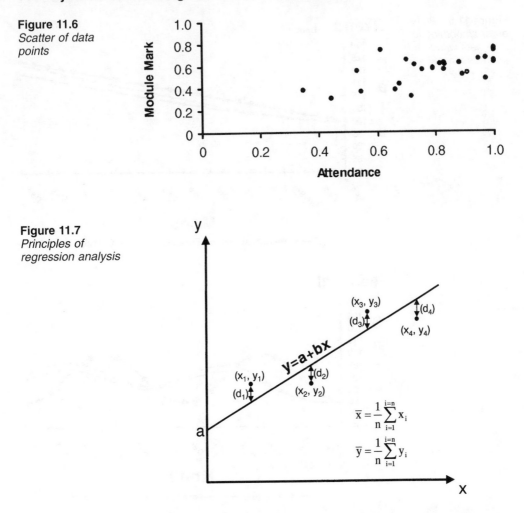

Figure 11.6
Scatter of data points

Figure 11.7
Principles of regression analysis

obvious questions present themselves. First, is there a correlation between attendance and good performance? Second, if there is a relationship between performance and attendance, how can this be quantified? This problem can be solved by regression analysis.

The basic theory is illustrated in Figure 11.7. If there is a scatter of points, (x_i, y_i), the purpose of regression analysis is to find the best line, $y = a + bx$. The derivation of the basic equations is beyond the scope of this book; however, it can be shown that the gradient of the line, b, that minimizes the total distance Σd_i is given by equation (11.5). The intercept, a, is given by equation (11.6).

$$b = \frac{n\Sigma xy - \Sigma x \Sigma y}{n\Sigma x^2 - (\Sigma x)^2} \tag{11.5}$$

$$a = \bar{y} - b\bar{x} \tag{11.6}$$

To understand the technique (sometimes also called *least squares*), consider Table 11.1. This shows the base data used to plot the graph in

Table 11.1 Regression example

Person	Att	Mark	xy	x^2
A	0.35	0.38	0.13	0.12
B	0.44	0.31	0.14	0.19
C	0.53	0.55	0.29	0.28
D	0.54	0.36	0.20	0.30
E	0.61	0.74	0.45	0.37
F	0.66	0.38	0.25	0.44
G	0.67	0.43	0.29	0.45
H	0.70	0.65	0.46	0.49
I	0.71	0.32	0.23	0.51
J	0.73	0.60	0.44	0.53
K	0.75	0.56	0.42	0.57
L	0.79	0.57	0.45	0.63
M	0.79	0.57	0.45	0.63
N	0.82	0.61	0.50	0.67
O	0.83	0.56	0.46	0.69
P	0.83	0.62	0.52	0.69
Q	0.83	0.60	0.50	0.69
R	0.88	0.62	0.55	0.78
S	0.89	0.51	0.46	0.80
T	0.91	0.53	0.48	0.83
U	0.95	0.65	0.62	0.90
V	0.97	0.48	0.47	0.95
W	0.97	0.66	0.65	0.95
X	1.00	0.75	0.75	1.00
Y	1.00	0.75	0.75	1.00
Z	1.00	0.64	0.64	1.00
A1	1.00	0.62	0.62	1.00
B1	1.00	0.73	0.73	1.00
28	22.18	15.79	12.91	18.44

Figure 11.8 *Line of best fit*

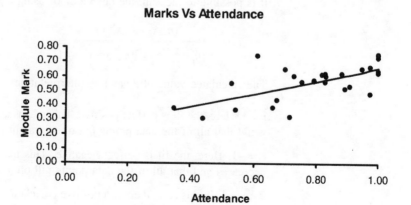

Marks Vs Attendance

Figure 11.8 and also the values required to be substituted into equation (11.5). Substituting these values, the values of *a* and *b* can be calculated.

The parameters have been used to plot the line of best fit in Figure 11.8.

$$b = \frac{(28 \times 12.91) - (22.18 \times 15.79)}{(28 \times 18.44) - (22.18)^2}$$

$$b = 0.454$$

$$a = \bar{y} - b\bar{x}$$

$$a = \left[\frac{1579}{28}\right] - 0.454 \times \left[\frac{2218}{28}\right]$$

$$a = 0.204$$

When considering extrapolation in manufacturing forecasting, correlation can arise in three ways:

- **Inside the company.** In this case, statistical information from one business area is used in another. The commonest example of this is the forecasting of spares demand. For automotive components, demand for aftermarket items is related to the number of items sold as original equipment in the past.
- **Inside the industry.** Here, statistical information of the general manufacturing sector can be used. For example, in the automotive manufacturing components sector, the number of aftermarket items sold is related to the number of new vehicles sold in the past.
- **Outside the industry.** This normally refers to the influence of general economic conditions to the product demand. For example, large car sales are significantly influenced by oil prices.

11.4.3 Correlation coefficients

It is possible to measure the degree of fit using equation (11.7).

$$r = \frac{(n\sum xy) - (\sum x \sum y)}{\sqrt{[n\sum x^2 - (\sum x)^2][n\sum y^2 - (\sum y)^2]}} \qquad (11.7)$$

The calculated value of r has five interpretations:

$r = 1$. Here, the fit is perfect positive. This means as x increases, so does y and that all of the data points fall on a straight line.

$r = -1$. Here, the fit is perfect negative. This means as x increases, y decreases and that all of the data points fall on a straight line.

$0 < r < 1$. In this case, there is a positive correlation, but all of the points do not fall onto a straight line.

$0 > r > -1$. In this case, there is a negative correlation, but again all of the points do not fall onto a straight line.

$R = 0$. No correlation.

These cases can be represented diagrammatically as shown in Figure 11.9.

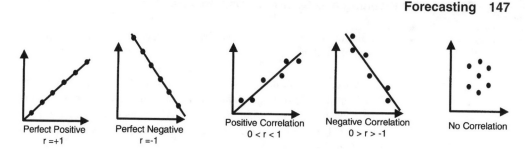

Figure 11.9 *Diagrammatic representation of correlation coefficient*

11.4.4 Multiple regression

In practice, a forecasted variable may be dependent on two or more variables. For example, house prices in the near future will be influenced by many economic parameters: interest rates, levels of unemployment, etc. Similarly, to forecast the mark of a student, it is sensible to consider previous marks as well as attendance.

In practice, multiple regression analysis can only be carried out using computers. Details of multiple regression are beyond the scope of this book. There are, however, computer packages (such as the Statistical Package for the Social Sciences or SPSS) that are available for multiple regression analysis.

11.5 Forecast uncertainty

All forecasts have one thing in common: they are subject to uncertainty. In practice, all companies have to acknowledge and plan around this fundamental truth. Details of how safety stocks can be used to compensate for uncertainty in demand in the context of inventory planning is presented in part 16.6.

11.6 Improving forecast accuracy

Accurate forecasts are essential for manufacturing firms. There is a limit to what can be achieved by purely statistical means. There are, however, two strategies that can be employed to improve accuracy:

Reduce lead time. If a forecast is made of events occurring in the distant future, it is inevitable that the forecast will be inaccurate. By reducing lead times and making the organization more responsive, forecasts need to be made only over shorter periods.

Aggregate forecast. Forecasts at an aggregated level are more accurate that those where the data is considered independently. For example, it is difficult to forecast the height of a person drawn at random from the population. It is much easier to forecast the *average* height of ten people drawn at random.

The reason for this improved accuracy lies in the nature of statistics. Consider the case of four products being manufactured from a different casting. The demand variation for the four products would be directly transmitted to the demand for the castings. This is shown in the lower part of the graph of Figure 11.10.

Figure 11.10
Aggregate demand patterns

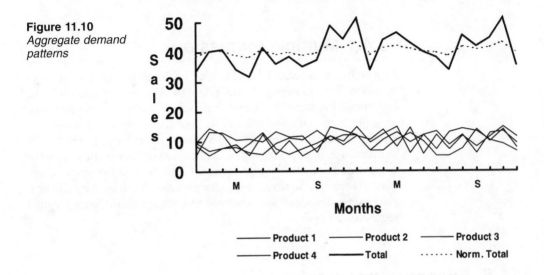

If the design was changed, however, such that the four items were made from a single casting, demand variation would be 'damped out'. The dashed line at the top of Figure 11.10 shows the aggregate demand. This aggregate demand represents four times the demand. The solid line at the top of Figure 11.10 shows the normalized demand for the casting (i.e., the variation divided by four). It can be seen that the aggregate demand has less variation than any of the four components. This can be explained mathematically. If there are n data sets each with a mean value, the sum of means gives the aggregate mean:

$$\bar{x}_a = \sum_{i=1}^{i=n} \bar{x}_i \tag{11.8}$$

The aggregate standard deviation, on the other hand, is given by the sum of squares:

$$\sigma_a^2 = \sum_{i=1}^{i=n} \sigma_i^2 \tag{11.9}$$

From equations (11.8) and (11.9), it is possible to show that equation (11.10) is true:

$$\frac{\sigma_a}{\bar{x}_a} < \sum_{i=1}^{i=n} \frac{\sigma_i}{\bar{x}_i} \tag{11.10}$$

Forecasting summary

Forecasting is important for all manufacturing businesses. There are three basic methods of forecasting, all with their respective advantages. It must be recognized, however, that forecasting is subject to inherent uncertainty. It is necessary, therefore, to take appropriate steps to compensate for uncertainty. Finally, there are techniques for reducing forecast error by changing the nature of the problem.

12 Optimization

12.1 Overview

One of the commonest problems to be addressed in the general area of QDA is optimization. An optimum is defined as follows:

The most favourable conditions; the best compromise between opposing tendencies; the best or most favourable.

This chapter will review different optimization techniques that are applied in manufacturing.

12.2 The objective function

Before discussing optimization methods, it is necessary to introduce *objective functions*. This is a precise, quantitative definition of the goal of a problem. In a given problem situtation, there might be several legitimate objective functions.

Defining the objective function is the essential first step in undertaking an optimization exercise. While this seems a simple task, the choice of an objective function is often neither obvious nor easy. Consider the following list of objectives:

1 Maximize profit.
2 Maximize cashflow.
3 Maximize sales.

All of these are legitimate objectives for a manufacturing organization. For example, in the long term a company may wish to maximize profits. The commonest reason that businesses fail, however, is not poor profitability, but poor cashflow. Often, companies that are essentially sound simply run out of money. For this reason, under adverse trading conditions, it is not unusual for companies to adopt a strategy of maximizing cashflow. This can be achieved in several ways; selling off slow moving stocks at a discount, for example. This can, however, have adverse effects on profit. Alternatively, if a company is moving into a new market, it may be desirable to seize a substantial market share. This could be achieved by setting low selling prices. This will of course affect both profit and cashflow.

Setting an objective function is therefore a strategic issue based on judgement. Once the objective function has been established, however, the process of optimization is analytical and quantitative.

Figure 12.1 *Plot of function f(x)*

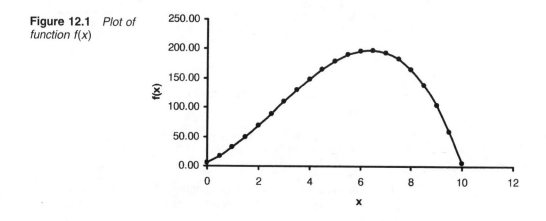

12.3 Basic optimization concepts

The simplest optimization case is that of a function of a single variable. Consider the function below, plotted in Figure 12.1:

$$f(x) = -x^3 + 8x^2 + 20x + 5$$

By inspection, the optimum lies at $x.6.4$. If $f(x)$ is an objective function, finding the optimum value in the range $0 < x < 10$ can be determined by algebraic means. The method is first to differentiate the function:

$$\text{By } f'(x) = -3x^2 + 16x + 20 = 0,$$

setting the derivative to zero the point(s) of inflexion in the defined range can be determined by solution of the equation. Using the standard formula for a quadratic equation:

For $y = ax^2 + bx + c$

$$\text{Quadratic roots} = \frac{-b \pm \sqrt{b^2 - 4ac}}{2a} \qquad (12.1)$$

By substitution, it can be seen that the optimum value in the nominated range is 6.38. Difficulties arise, however, if the function cannot readily

Figure 12.2 *Plot of*
f(x), 0 < x < 10

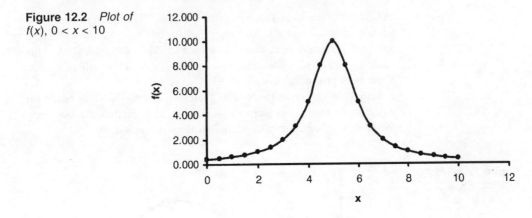

be differentiated. For example, consider the formula below, plotted in
Figure 12.2:

$$f(x) = \frac{10}{(5 - x)^2 + 1}$$

This function cannot be readily differentiated and set to zero (the chain
rule needs to be applied). By inspection, however, the optimum occurs
when x.5.

Problems of this type can be solved numerically. That is to say that *all*
points are evaluated and the largest value selected. The points used to
plot the function f(x) in Figure 12.2 are shown in Table 12.1.

From Table 12.1, it can be seen that the optimum lies in the range 4.5
< x < 5.5. Thus f(x) data can be evaluated in this range and plotted once
more.

In principle this numerical approach can be used to solve problems
with an indefinite number of variables. Consider the function below:

$$f(x,y) = -2x^3 + 20x^2 + 20x - 2y^3 + 20y^2 + 20y$$

The optimum value in the range 0 < x < 10, 0 < y < 10 can be established
by homing in on the optimum point as in the single variable problem.
This is shown in Figure 12.4.

Table 12.1 Values of f(x) for 0 < x < 10

x	0.00	0.50	1.00	1.50	2.00	2.50	3.00	3.50	4.00	4.50	5.00	5.50	6.00	6.50	7.00	7.50	8.00	8.50	9.00	9.50	10.00
f(x)	0.38	0.47	0.59	0.75	1.00	1.38	2.00	3.08	5.00	8.00	10.00	8.00	5.00	3.08	2.00	1.38	1.00	0.75	0.59	0.47	0.38

OPT

Table 12.2 Values of f(x) for 4.5 < x < 5.5

x	4.50	4.55	4.60	4.65	4.70	4.75	4.80	4.85	4.90	4.95	5.00	5.05	5.10	5.15	5.20	5.25	5.30	5.35	5.40	5.45	5.50
f(x)	8.00	8.32	8.62	8.91	9.17	9.41	9.62	9.78	9.90	9.98	10.00	9.98	9.90	9.78	9.62	9.41	9.17	8.91	8.62	8.32	8.00

OPT

Figure 12.3 *Plot of f(x), 4.5 < x < 5.5*

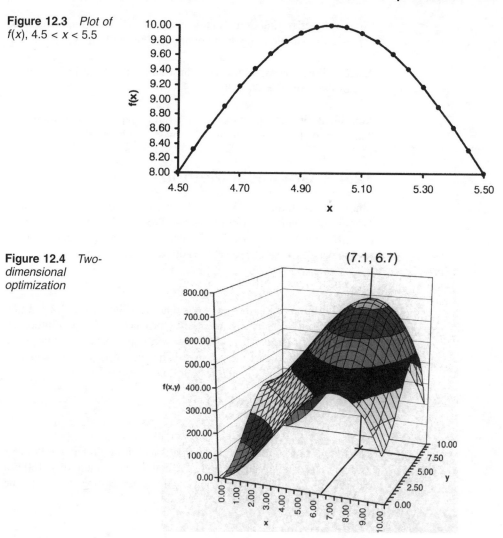

Figure 12.4 *Two-dimensional optimization*

Clearly, manual calculation is impractical in these cases and a computer must be used. Spreadsheets can be highly effective in this class of problem.

12.4 Linear programming

12.4.1 Basic concepts

Another common problem is that of optimization of variables under conditions of constraint. The mathematical space that is free of constraint is called the *feasible space* or *feasible area*. A common technique used to tackle problems of this sort is *linear programming*. All linear programming problems have three characteristics:

Linear objective function. That is to say, the objective function can be expressed in simple terms in an expression of the form $K = Ax + By$.

A set of linear constraints. The constraints on the feasible space are also of the form $K = Ax + By$.

Non-negativity. The problem is solved in a graphical space where $x > 0$ and $y > 0$. Thus optimum answers may only be positive.

12.4.2 Graphical method

Linear programming is best understood by considering a simple example. A manufacturer produces two products X and Y. X is made of cast iron and Y is an alloy. The company's objective function is sales revenue and this is defined by the following expression:

Sales $= 1.6X + 1.4Y$

The implication of this expression is that the selling price of X is £1.60 and Y is £1.40. Ideally, the number of products of X and Y produced would be very large. There are, however, constraints. The first is the power available to cast the products. The second is machine capacity. These constraints are given by the following expressions:

Power $\rightarrow 160 > 1.34X + Y$

Machining capacity $\rightarrow 150 > X + 1.25Y$

The implication of the first expression is as follows: it takes 1.34 times as much power to produce a unit of X compared to a unit of Y. 160 is a measure of the total power being available for casting. Similarly, the

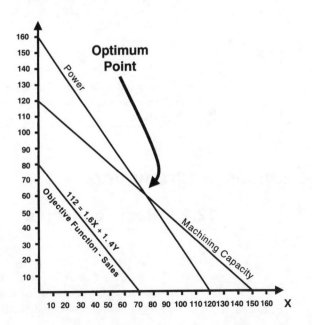

Figure 12.5 *Linear programming – two constraints*

implication of the second expression is that it is 1.25 times more time consuming to machine alloy components (Y) than cast iron components (X). 150 is a measure of the total time available for machining. This information can be represented graphical as shown in Figure 12.5.

Finding the optimum point in this simple case can be achieved by the simultaneous solution of the two constraint expressions:

$$160 = 1.34X + Y$$

$$150 = X + 1.25Y$$

$$\therefore \quad X = 150 - 1.25Y$$

$$160 = 1.34(150 - 1.25Y) + Y$$

$$160 = 200 - 1.667Y + Y$$

$$160 = 200 - 0.667Y$$

$$0.667Y = 200 - 160$$

$$Y = \frac{200 - 160}{0.667} = 60$$

$$\therefore \quad X = 150 - (1.25 \times 60) = 75$$

The problem becomes far more complex, however, if additional constraints are added. Consider a third constraint of labour capacity given by the following expression:

$$\text{Labour cap} \rightarrow 130 > X + Y$$

This information is shown in Figure 12.6.

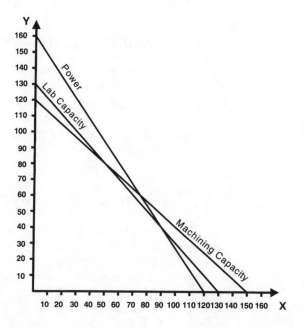

Figure 12.6 *Linear programming – three constraints*

Figure 12.7
Solution to the problem

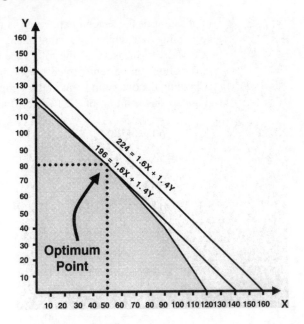

The three constraints define a feasible region. To find the optimum point graphically, the sales line needs to move down until it touches the feasible region. This is shown in Figure 12.7.

Reading from the graph, the optimum sales value will be achieved if $50X$ and $80Y$ are produced.

Algorithms exist to generate the optimum values (the most common being the Simplex method). These are, however, beyond the scope of this book. With a larger number of constraints, these algorithms are the only practical way of solving this type of problem.

12.4.3 Multiple variables

So far, only problems with two variables have been discussed. In practice, an optimization problem may have an indefinite number of variables. These problems are difficult to visualize and still harder to solve. As before, these problems can only be solved in practice using algorithms (for example, the well-known Simplex method) running on powerful computers.

12.4.4 Sensitivity analysis

All of the examples used so far have assumed the information used in the analysis is accurate and deterministic. In practice, however, information is imperfect; that is to say inaccurate and stochastic. It is often useful, therefore, to conduct *sensitivity analysis*.

Sensitivity analysis involves re-examining a problem to determine what effect a change in the input variables will have on the final answer.

Figure 12.8
*Illustration of
sensitivity*

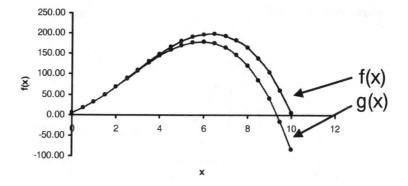

Consider the initial function in Figure 12.1. Figure 12.8 shows the effect
of a small change in the coefficients in the original function:

$$g(x) = -1.1x^3 + 8.1x^2 + 20.1x + 5$$

There are also cases where small changes in the input parameters have
a large effect. A particular case is that of profit and loss models.

Profit is the difference between revenue and cost. If costs are fixed,
small increases or decreases in sales will have a large effect on profit.
Figure 12.9 shows how return on sales (ROS) varies with sales volume
(ROS = Profit/Sales, where Profit = Sales – Costs). In the example,
costs are £1000 K.

Other models are less critical. For example, optimum lot size models
are relatively insensitive. Small changes in lot size have a minimal
effect on costs (see Figure 12.10).

By examining the model's tolerance to changes, it is possible to gain
an impression of its sensitivity. In many cases, a model can be very
sensitive to one parameter, but very insensitive to another.

When undertaking a sensitivity analysis, one or both the following
may be undertaken:

- **Test the sensitivity of the model itself.** Here, the analyst will
  investigate how sensitive the outcome of a model is with reference
  to its own structure. Typically, the analyst will undertake tests with

Figure 12.9 *Effect
of sales on ROS*

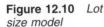

Figure 12.10   *Lot size model*

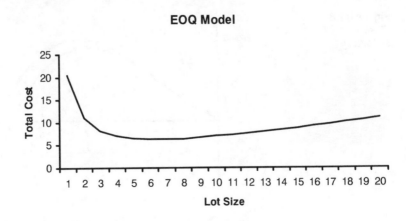

a particular model. The analyst will then slightly modify the parameters governing the model and rerun the tests. By comparing the outputs from the two sets of tests (as in Figure 12.8), the analyst can form an impression of the sensitivity of the model.

● **Test the sensitivity of the model to input variables.** In this case, the analyst will examine the behaviour of the model over a range of operations. For example, it can be seen from Figure 12.9 that return on sales (ROS) is very sensitive to small changes in sales volume.

Sensitivity analysis is a useful technique for optimization problems, but it is also applicable to checking the robustness of any model.

12.5 Optimization summary

Optimization is a valuable technique for the practising manager. It must be recognized, however, that it is dependent on the identification of a single objective function. In practice, a balance must be struck between conflicting and often mutually exclusive objectives. Because of this, optimization is only a support tool for management. It cannot completely replace human judgement.

13 Evaluating alternatives

13.1 Overview

One of the primary roles of the manufacturing manager is to evaluate alternatives and choose the most appropriate. Ultimately, this process is the responsibility of the manager concerned and will involve personal judgement. It is desirable, however, that the evaluation process is assisted by quantitative techniques.

This chapter will focus on *decision tree*. This is a technique for systematically structuring the decision-making process and placing numerical values on the benefits (usually financial) of different outcomes.

13.2 Decision theory

13.2.1 Overview

The basic approach for any systematic decision process is as shown in Figure 13.1.

This is best explained using an example. A student living in student accommodation is considering taking on a mortgage. The rationale for this is financial; at the end of the period of study, the student concerned would have a property that could be sold. Thus, applying the decision-making process above:

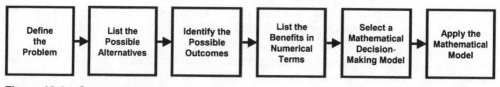

Figure 13.1 *Systematic decision process*

1 **Define the problem.** The problem is deciding whether or not a property should be purchased.

2 **List the alternatives.** In this case, there are three alternatives: buy a house, buy a flat or remain in student accommodation (i.e. do nothing).

3 **Identify the possible outcomes.** There are several outcomes of this decision-making process. There is some risk associated with purchasing a property (interest rates, house/flat prices, etc.). In this case, it can be assumed that there are two possible sets of conditions, favourable (where buying property is beneficial) and unfavourable (where renting property is more advantageous). These different outcomes are referred to as *states of natures* (as the decision maker has little or no control over these events).

4 **List the benefits in numerical terms.** This can be summarized in Table 13.1.

5 & 6 **Select and apply mathematical decision-making model.** The final stage is to select/apply a mathematical decision-making model. The model selection will depend on the environment.

Table 13.1   Decision table

Alternatives	States of Nature	
	Favourable	Unfavourable
House	£10 000	−£2000
Flat	£6000	−£1000
Rent	£0	£0

13.2.2 Decision-making situations

There are three different types of decision-making situations:

● **Decision making under certainty.** In this case, all of the consequences are known. For example, two building society accounts with fixed three-year terms can be compared and evaluated without uncertainty. Typically, in manufacturing, such circumstances of certainty are rare. In the example in Table 13.1, if it was certain that conditions would be favourable, then the decision would be clear; the student should buy a house.

● **Decision making under uncertainty.** In this case, no information is available about the relative likelihood of the different outcomes. In this case, judgement or some rule of thumb must be applied in the absence of hard information.

● **Decision making under risk.** Here, information exists on the probabilities of the different outcomes. This allows a systematic approach to be taken to the decision-making process.

13.2.3 Decision making under risk

If we assume there is a 0.75 probability of favourable conditions for buying property, it is possible to draw the decision tree shown in Figure 13.2.

Figure 13.2
Decision tree

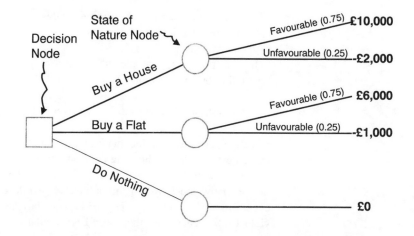

To evaluate the different possibilities, it is necessary to introduce the idea of expected monetary value (EMV). If the decision-making processes could be undertaken a large number of times, the EMV would be the average value of each possible outcome. EMV is defined as follows:

$$\text{EMV}(x) = \sum_{i=1}^{i=n} x_i.P(x_i) \tag{13.1}$$

Using equation (13.1), it is possible to calculate the EMV value of each of the decisions:

$$\boxed{\text{EMV (house)} = (0.75 \times £10\,000) + (0.25 \times -£2000) = £7000}$$

$$\text{EMV (flat)} = (0.75 \times £6000) + (0.25 \times -£1000) = £4250$$

$$\text{EMV (nothing)} = (1.0 \times £0) = £0$$

The logical decision based on the information would suggest that the student should buy a house. Note that it is possible to handle decision problems with more than two states of nature and multiple stages.

13.2.4 Value of perfect information

In many decisions, studies or research can improve the quality of the information available. Unfortunately, this will involve cost and implies a further decision; is it worth investing money in obtaining better information? To address this problem, the concept of *expected value of perfect information* (EVOPI) was formulated. To understand this concept, it is first necessary to appreciate the concept of *expected value* with *perfect information* (EVWPI). It is defined as follows:

$$EVWPI = \sum_{j=1}^{j=m} x_{bj}.P(x_j) \qquad (13.2)$$

where m = the number of different states of nature, x_{bj} = the best outcome for a particular state of nature j and $P(x_{bj})$ = the probability of the state of nature j.

In the previous example, there are only two states of nature, favourable or unfavourable. Thus the EVWPI is as follows:

$$EVWPI = [£10\,000 \times 0.75] + [£0 \times 0.25] = £7500$$

Effectively, this is the value that might be expected by avoiding negative outcomes; i.e. always choosing the best option. Consider the case of the student seeking professional advice as to the future state of the property market. How much would this advice be worth, even in the unlikely event this was perfect? The best EMV is of buying a house, which will yield £7500. Therefore, the EVOPI is calculated as follows:

$$EVOPI = EVWPI - \text{Maximum EMV}$$
$$EVOPI = £7500 - £7000 = £500 \qquad (13.3)$$

Therefore, the most that any advice (*even* if perfect) would be worth is £500.

It should be noted, however, that the analysis above simply provides support to the decision-making process. Inevitably, the student concerned will need to consider his/her personal circumstances and the level of risk s/he is prepared to accept. Ultimately, the decision is *qualitative*, albeit supported by *quantitative* analysis.

13.2.5 Decision making under uncertainty

It is common for managers to have to make decisions with minimal information (i.e. no information of relative probabilities exists). In this case, less systematic criteria must be applied. Several different criteria exist, but in this book only three will be discussed.

- **Maximax.** In this case, the selected alternative will be the maximum outcome for every alternative. This is achieved by finding the maximum outcome within every alternative and then selecting the highest value in the row of the decision table. This represents the highest possible gain and is therefore sometimes called the *optimistic decision criteria*.
- **Maximin.** Here, the minimum outcome for every alternative is found and the maximum of these is selected. This represents the minimum lost and is therefore sometimes called the *pessimistic decision criteria*.

Figure 13.3
Decision-making under uncertainty

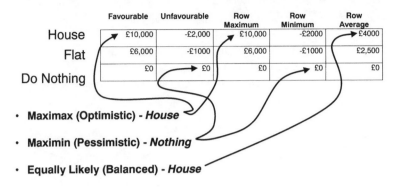

	Favourable	Unfavourable	Row Maximum	Row Minimum	Row Average
House	£10,000	-£2,000	£10,000	-£2000	£4000
Flat	£6,000	-£1000	£6,000	-£1000	£2,500
Do Nothing	£0	£0	£0	£0	£0

- **Maximax (Optimistic) - *House***
- **Maximin (Pessimistic) - *Nothing***
- **Equally Likely (Balanced) - *House***

- ● **Equally likely.** This starts from the arbitrary premise that all of the states are equally likely. The average outcome for each alternative is calculated. The alternative selected is the maximum of the average values.

This is summarized in Figure 13.3.

13.3 Decision-making example

This example is based on the problem outlined at the start of this part. Should the two sisters take their car or take a chance on getting a lift? Summarizing the problem:

Two sisters plan to go to a party. A lift is available to get to the party, but there is no guarantee this will be available for the return journey. It may be necessary to use a taxi therefore and this will cost £7.50. Taking the car will cost £2.00.

- ● Produce a decision table for this case.
- ● What is the best decision based on the maximax, maximin and equally likely criteria?
- ● The probability of obtaining a lift is 0.4. Draw a decision tree for this case.
- ● Should the sisters take the car based on the probability information given?

The decision table is as shown below:

Alternatives	States of nature				
	Lift	No Lift	Row Max	Row min	Row average
Take car	-£2.00	-£2.00	-£2.00	-£2.00	-£2.00
Leave car	£0.00	-£7.50	-£0.00	-£7.50	-£3.75

- Maximax = leave car
- Maximin = take car
- Equally likely = take car

The information can also be presented on a decision tree.

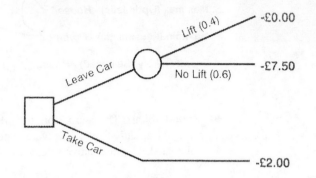

Calculating the EMV for the two alternatives:

EMV (take car) = –£2.00

EMV (leave car) = (0.4 × £0.00) + (0.6 × –£7.50) = –£4.50

The best alternative therefore is to take the car. It should be noted, however, that the decision cannot be evaluated on purely financial grounds. Choosing to take their own car will mean one of the sisters will not be able to have an alcoholic drink.

14 Simulation

14.1 Overview

Simulation is defined as follows:

The imitation of a real world system.

Examples of simulation include the following:

- Flow of people through an airport.
- Nuclear war.
- Traffic flow in a large city.
- Behaviour of a suspension system.
- World economy.
- Temperature/stress in a piston.
- Forces in a metal cutting process.
- Queues in a manufacturing system.

There are a number of reasons why simulation is employed:

Practicality. With large systems, simulation may be the only practical way of approaching the problem.

Safety. Simulation may allow a system to be tested under extreme conditions that could not be justified in the real world.

Allows 'what-if' analysis. This allows the physical system to be built 'right first time'. This has implications for cost/time to implement the real world system. Also, simulation is repeatable.

Promotes understanding. Simulation can provide a better understanding of systems, particularly if allied to graphics. Simulation can also verify analytical solutions to problems.

As computer hardware has become cheaper and more powerful, simulation has become an increasingly common engineering tool. It is now feasible to employ graphical animation techniques which require enormous rates of computation. Virtual reality (VR) is a form of simulation that is perhaps attracting the greatest interest at present. It should be noted that manual simulation is possible, though with PCs readily available, this is now fairly rare.

14.2 Terminology

Before starting a study of simulation, it is important to understand the terminology used:

- **System environment.** This is the 'context' in which the simulation takes place.
- **System boundary.** The dividing line between the system and the external environment.
- **Entity.** A component of the system.
- **Attribute.** A property of an entity.
- **State.** A collection of variables to define the condition of an entity.
- **Endogenous event.** A change in state arising from *inside* the system.
- **Exogenous event.** A change in state arising from *outside* the system.
- **Model.** A formal representation of the system environment.

14.3 Simulation methodology

It will be seen later in this chapter that there are several techniques for simulation, each suited to meet particular objectives. There is, however, a methodology that can be applied to all simulation projects:

Step 1 Build a conceptual model of the system to be simulated.
Step 2 Convert the conceptual model into a formal model that can be entered into the computer system.
Step 3 Verify the model.
Step 4 Undertake a series of experiments with the model.
Step 5 Draw conclusions about the system.

It should be noted that the objectives of the simulation would have a large effect on the process. For example, it is unlikely that cosmetic appearance will be a factor in the simulation of a factory. The same is not true, however, of the simulation of an airport.

Table 14.1 Hierarchy of simulation techniques

	Continuous	Kinematic	Robotic
	Hybrid		
Simulation	Discrete	Static	
		Dynamic	Deterministic
			Stochastic

14.3.1 Types of simulation techniques

There is a hierarchy of simulation techniques (Table 14.1).

Most of this chapter will be devoted to discrete event simulation, though it is important to introduce the concepts of continuous simulation.

14.4 Continuous simulation

The two principal divisions of simulation are continuous and discrete, though hybrids of the two can arise.

Continuous simulation is concerned with the behaviour of a system over time. It is usually possible to describe such systems by means of differential equations. These differential equations can be solved to yield a model of the system as shown in the example in Figure 14.1.

If these equations are non-linear and/or complex, an analytical solution may not be possible. One particular type of this simulation that is relevant to manufacturing engineering is kinematic. This technique uses 'primitive' mechanisms (illustrated in Figure 14.2) as entities to

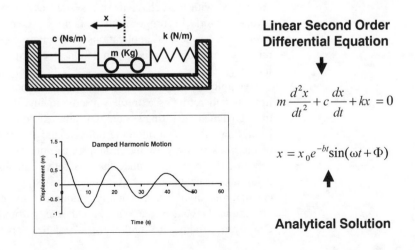

Linear Second Order Differential Equation

$$m\frac{d^2x}{dt^2} + c\frac{dx}{dt} + kx = 0$$

$$x = x_0 e^{-bt}\sin(\omega t + \Phi)$$

Analytical Solution

Figure 14.1
Continuous simulation

Figure 14.2
Primitive mechanisms

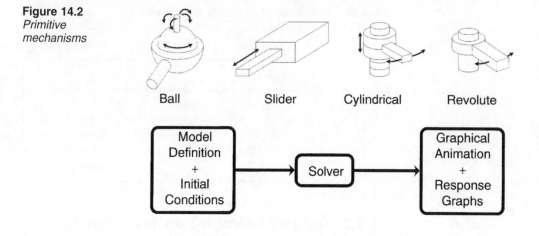

Ball   Slider   Cylindrical   Revolute

create a representation of a complex mechanical system. Packages such as 'GRASP' can be applied to model the motion of such systems. Some sophisticated modellers allow each entity to be assigned attributes such as mass and moment of inertia and also permit the addition of non-rigid elements such as spring-dampers. The output of such a simulation is usually in the form of graphical animation, though more detailed quantitative data is also available. Some CAD packages such as 'CATIA' and 'I-DEAS' have integrated kinematic simulation systems. One particular area where this technique can be applied is robotics.

14.5 Discrete event simulation (DES)

DES is applied widely in manufacturing. Where continuous simulation deals with continuous variables (such as displacement, velocity or acceleration), DES deals with discrete events such as whether a coin is a head or a tail or a machine is operating or not. This can be further divided into static and dynamic simulation. Static simulation is used to model an environment independent of time. Dynamic simulation is concerned with the behaviour of a system over time (for example, the level of work in progress in a factory). Dynamic discrete event simulation has several applications in a variety of fields. It is also the most commonly applied simulation technique in manufacturing industry.

Dynamic simulation is concerned with the behaviour of a system over time and therefore is extremely useful in investigating material flow through a manufacturing system. Dynamic discrete event simulation can be further subdivided into deterministic and stochastic types. This part of the book will review the principles of both of these types.

While it is possible to undertake discrete event simulation by hand, it is far more usual to employ a computer due to the laborious nature of the calculations involved. Originally, standard high-level computer programming languages such as Pascal and particularly Fortran were employed. Today, a number of specialized packages have been developed and are now commercially available (for example, 'HOCUS' and 'ProModel').

14.5.1 Deterministic simulation

Deterministic processes are those that are repeatable. This contrasts with stochastic processes which can only be described by a probability distribution. In a deterministic model, there is no uncertainty in any system event. For example, in a factory simulation, the processing time for a particular component at a given machine will be consistent and thus predictable.

An example of a deterministic simulation can be seen in Figure 14.3.

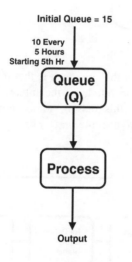

Time Period	Q	Event(s)
Start-up	15	
End of 1st Hour	15	
End of 2nd Hour	8	7 Units Used
End of 3rd Hour	8	
End of 4th Hour	8	
End of 5th Hour	11	7 Units Used, 10 Units Arrive
End of 6th Hour	11	
End of 7th Hour	11	
End of 8th Hour	4	7 Units Used
End of 9th Hour	4	
End of 10th Hour	14	10 Units Arrive
End of 11th Hour	7	7 Units Used
End of 12th Hour	7	
End of 13th Hour	7	
End of 14th Hour	0	7 Units Used
End of 15th Hour	10	10 Units Arrive
End of 16th Hour	10	
End of 17th Hour	3	7 Units Used
End of 18th Hour	3	
End of 19th Hour	3	
End of 20th Hour	10	3 Units Used, 10 Units Arrive
End of 21st Hour	10	
End of 22nd Hour	10	

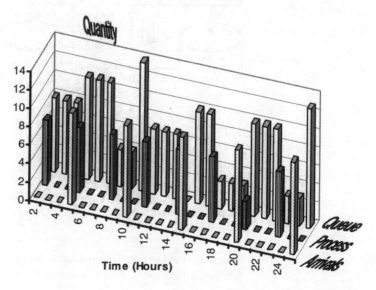

Figure 14.3 *An example of a deterministic discrete event simulation model*

14.5.2 Simulation methods

The simplest method for simulation is time slicing. Here, time is separated into a series of discrete buckets. This method is illustrated in the flow chart shown in Figure 14.4.

Figure 14.4   *Time slicing method for DES*

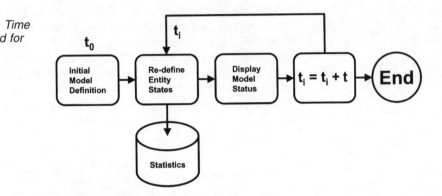

The main limitation of this approach is that if the time interval (t) is of the same order of magnitude as the interevent period, a distorted picture will result. For this reason, some commercial packages are more sophisticated and use the *next event* approach illustrated in Figure 14.5.

Figure 14.5   *Next event method for DES*

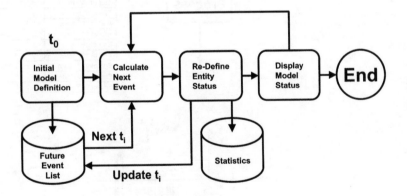

The stages are as follows:

Initial model definition. It is first necessary to define the entities (via their attributes) and the relationships between them. It is also necessary to define the initial states of each entity at time = t_0. The first stage also needs to define the future event list (FEL). This defines the timing of future events for all entities.

Calculate next event. The simulation model needs to define the time at which the next event, t_i, will occur.

Redefine model status. At the new time, t_i, the state of each entity in the model is redefined. This will also update the FEL. Statistics about the overall characteristics of the model will probably be collected.

Display model status. At the end of the cycle, it is possible to display the status of the entire model. In most modern simulation packages, this display will be in graphical form.

End. The simulation will return to the 'Redefine entity status' stage until a predefined end-point is reached.

14.5.3 Stochastic simulation

Many manufacturing processes are stochastic. For example, the time to complete a manufacturing operation may vary each time it is undertaken. It is possible, however, to build this variation into a simulation model. The output from this model will be different each time the program is run. General conclusions, such as the average queue length, will be fairly repeatable nonetheless. Note that the process and arrival times are (somewhat unrealistically) based on linear probability distribution functions.

A more realistic example is the well-known single channel queuing model as shown in Figure 14.6. This model uses Poisson and exponential PDFs for arrival and service respectively. The average queue length is dependent on the utilization. An analytical solution to this problem exists, though it is also possible to solve this case (and other more complex variants) via simulation.

Figure 14.6
Example – simple waiting line model

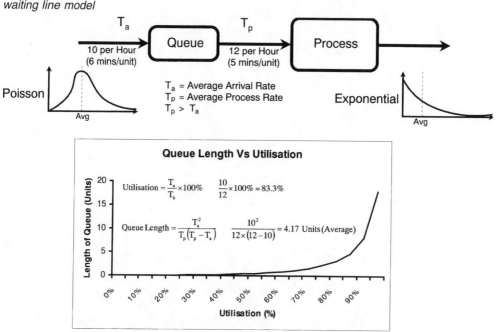

It should be noted that most modern simulation packages allow events other than arrival and service to be considered stochastically. Other events, such as machine set-up or breakdown, can also be represented in this way. It should also be noted that in order to build a realistic stochastic simulation model, it is necessary to understand the PDFs describing the model. In practice, this can be very difficult and time consuming.

14.6 Limitations of simulation

Despite the power of simulation, caution should be exercised in its use:

Simulations themselves can be costly. The definition and debugging of a simulation model can be difficult and time-consuming.

Base data can be difficult to collect. For example, determining a behaviour of a workcentre over time can be matter of guesswork.

Simulations are not reality. All simulation models depend on approximations and simplifying assumptions.

A false sense of security. With modern, sophisticated simulation packages, the results can be highly impressive, particularly if animated graphics are employed.

15 Project management

15.1 Overview

All engineers will be involved in managing a project at some stage in their careers. Simple projects can be managed informally, but with complex cases, formal methods need to be applied. While project management has been necessary for thousands of years, formal methods date back only to the mid-1950s. A formal project management technique allows tasks to be divided among a group. It also provides a mechanism for checking that projects are proceeding in a timely manner. This chapter will review techniques that can be applied for project management.*

15.2 Project networks

15.2.1 Definition of terms

Projects can be defined in terms of activities and events.

- An activity is something that must be completed as part of a project. In a building project, an activity might be 'construct frame'.
- Events are moments in time that are separated by activities. For example, following the activity 'construct frame', the corresponding event might be 'frame completed'.

Projects can be represented graphically using activities and events. There are two approaches as shown in Figure 15.1:

- **Activity on node (AON).** Here, activities are represented as nodes and events as arrows.

* Further information on project management can be found in *Business Skills for Engineers and Technologists* (ISBN 0 7506 5210 1), part of the IIE Text Book Series.

Figure 15.1
*Diagram
conventions*

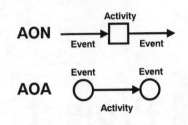

- **Activity on Arrow (AOA).** Here, activities are represented as arrows and events as nodes. This is an intuitive approach as events are linked by activities. The problem with this approach, however, as will be seen later, is that 'dummy' activities are needed to make diagrams logically consistent.

15.2.2 Network diagrams

Using the conventions above, network diagrams can be constructed. The first stage in any project planning exercise is to list the different tasks that might be required. To make a cup of the coffee for example, the following tasks are needed:

- Boil water (300 s)
- Locate coffee (60 s)
- Locate sugar (60 s)
- Locate milk (60 s)
- Add coffee and sugar (15 s)
- Add boiling water (15 s)
- Add milk (15 s)
- Serve coffee (120 s)
- Drink coffee (540s)

Note that the time required for each of these activities is shown in parenthesis. These tasks, however, are not independent. For example, 'Add boiling water' cannot take place until 'Boil water' has been completed. Example diagrams for making a cup of coffee are shown in Figures 15.2 and 15.3.

Note that the implication of Figures 15.2 and 15.3 is that the milk is only added after boiling water has been poured into the cup. Some

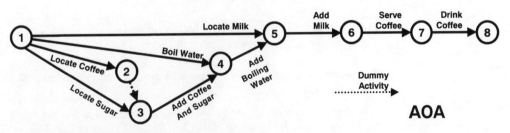

Figure 15.2 *AOA diagram for making coffee (1)*

Figure 15.3   *AON diagram for making coffee (1)*

Figure 15.4   *AOA diagram for making coffee (2)*

Figure 15.5   *AON diagram for making coffee (2)*

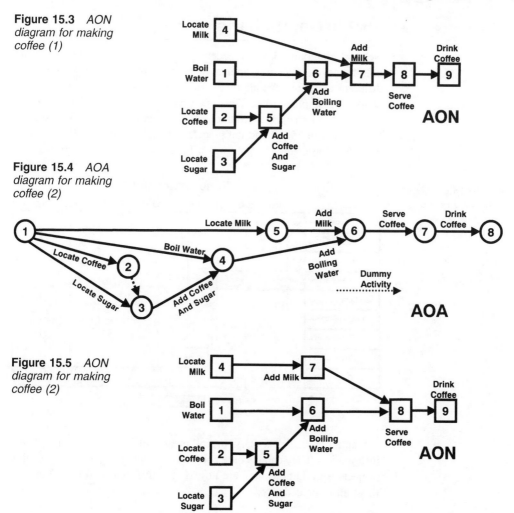

people, however, prefer to add the milk to the cup *before* the boiling water is poured onto the coffee and sugar in the cup. This change in precedence is shown in Figures 15.4 and 15.5.

Note also that the diagrams imply there are infinite resources. In this case, four people are assumed to be available to execute the first four activities.

Note that in Figures 15.2 and 15.4 there is a dummy activity represented by a dotted line. The activity 'Locate coffee' can be designated as (1–2). Without the dummy activity, however, the activity 'Locate sugar' would share the same designation, causing a logical inconsistency. Dummy events have zero duration.

The relative merits of the AOA and the AON are the subject of debate among project managers. In the past, for complex projects, AOA has generally been judged superior and has been more commonly applied. The leading software for project management employ the AON convention, however, and this means AOA is less commonly applied. This part of the book will therefore use the AON convention.

15.2.3 Gantt charts

Network diagrams show precedence of activities very clearly. They do not, however, give a graphical indication of the relative length of different activities. An alternative method of presentation is the Gantt chart. This approach was developed by Henry Gantt in 1914 (indeed, this was the first formal technique for project management). Using the times listed in 15.2.2 and the precedence logic defined in Figures 15.2 and 15.3, the Gantt chart shown in Figure 15.6 can be constructed.

Figure 15.6 *Gantt chart for making coffee*

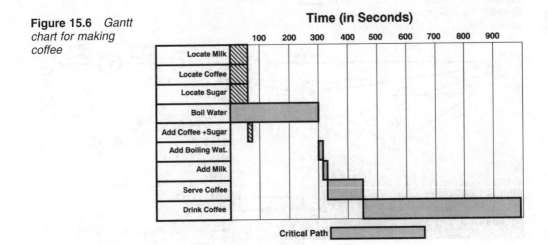

Gantt charts are a very useful means of communicating project information. They can become cumbersome, however, for very complex projects. In addition, they are not so effective as network diagrams in illustrating precedence.

15.3 Activity scheduling

15.3.1 Definitions

Note that in Figure 15.6, a *critical path* is indicated. A critical path is the longest path through the different activities in the project. The critical path therefore represents the shortest duration for the project. Those activities on the critical path are called *critical activities*. In the coffee-making example, it is clear that the first task to be undertaken by a single person is to boil the water. Note the following definitions:

Slack. The length of time available before an activity needs to start without delaying subsequent activities. In the coffee-making example, location of the milk does not need to start immediately.

Earliest start date (ESD). The earliest date an activity could start (determined by previous activities).

Earliest finish date (EFD). The earliest date that an activity could theoretically finish.

Latest start date (LSD). The latest date an activity could start without delaying subsequent activities.

Latest finish date (LFD). The latest date an activity could finish without extending the project.

15.3.2 ESDs, EFD, LSDs, LFDs and slack

EFDs are calculated in a forward pass through a network by adding the activity time (t_i) to the ESD as indicated in the following expression:

$$EFD = ESD + t_i \tag{15.1}$$

The ESD of a task is determined by the EFD of its immediate predecessor (thus initial tasks have an ESD of zero by definition). Where a task has two or more predecessors, its ESD is determined by the largest EFD of its predecessors.

Once all tasks have been given an ESD and an EFD, a backward pass through the network is used to calculate LSDs and LFDs. This is achieved using the following equation:

$$LSD = LFD - t_i \tag{15.2}$$

The LFD of a task is determined by the LSD of its immediate successor (thus final tasks have an ESD equal to the overall project time by definition). Where a task has two or more successors, its LFD is determined by the largest LSD of its successors.

Once all tasks have been given ESDs, EFDs, LSDs and LFDs, it is possible to calculate slack using the expression below (note that slack cannot be used more than once for tasks on the same path):

$$Slack = LSD - ESD = LFD - EFD \tag{15.3}$$

Tasks with zero slack are on the critical path. Figure 15.7 illustrates how the process is undertaken for the example described previously for making a cup of coffee. Note the format of the diagram: this will be used as the standard way of representing project networks in this book.

15.4 Project uncertainty

15.4.1 Overview

In practice, the duration of project activities are stochastic; i.e. they cannot be described by a single, deterministic value. A common

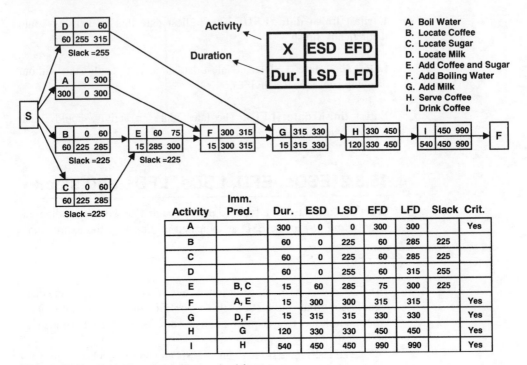

Figure 15.7 *Project network in standard form*

question is, therefore, what is the probability of a project being completed by a given deadline?

15.4.2 Analysis method

There is a six-stage process for determining the chances of a project being completed on time.

1 **Determine the most likely project time.** It is common practice to present a task as a β distribution (this is largely tradition). A β distribution is described by three parameters: a lower bound (a), an upper bound (b) and a most commonly occurring value (m). Distributions may be symmetrical or skewed as shown in Figure 15.8. The expected time (t_i) of a task i can be calculated using equation 15.4.

$$t_i = \frac{a + 4m + b}{6} \tag{15.4}$$

2 **Determine critical path and overall project slack.** Once the individual task times have been calculated, the critical path and the overall deterministic project length can be determined as described in 15.3. Table 15.1 summarizes the information related to the previous example for making a cup of coffee.

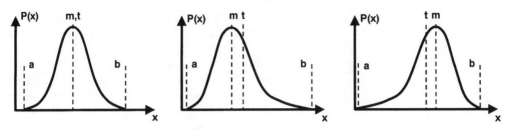

Figure 15.8 *Examples of β distributions*

Table 15.1 Data relating to coffee-making example

Activity i	a	m	b	t_i	Critical Y/N?	Critical Times	σ_i^2
A	240	300	360	300	Y	300	400
B	50	60	70	60	N		
C	50	60	70	60	N		
D	50	60	70	60	N		
F	12	15	18	15	Y	15	1
G	12	15	18	15	Y	15	1
H	96	120	144	120	Y	120	64
I	432	540	648	540	Y	540	1296
Total				1170		990	1762
						$\sigma_t =$	**42.0**

3 **Calculate total project time.** After the project has been analysed, non-critical items can be ignored. Thus, in the example, only tasks A, F, G, H and I need to be considered. The total length of the project (t_t) is given by the sum of these critical tasks (see equation (15.5)).

$$t_i = \sum_{i=1}^{i=n} t_i \tag{15.5}$$

4 **Calculate variances and total variance.** For a β distribution, the individual variances (σ_i^2) are given by equation (15.6):

$$\sigma_T^2 = \sum_{i=1}^{i=n} \sigma_i^2 \tag{15.6}$$

The central limit theorem states that if there are a large number of independent, random tasks, then the aggregate distribution will be normal (or Gaussian). The total variance for the critical path (σ_T^2) is given by equation (15.7).

$$\sigma_i^2 = \left(\frac{b-a}{6} \right)^2 \tag{15.7}$$

5 **Express project slack in terms of Φ.** Assume that the deadline for the example project is 17 minutes (1020 seconds). The total length of the critical path is 990, thus giving a project slack of 30 seconds.

Applying equations (15.6) and (15.7) yields a σ_T value of 42 seconds. Expressing this as a multiple of σ_T (see Table 15.1) 42 seconds = $30/42\sigma_T$ or $0.715\sigma_T$.

6 **Determine probability of σ_T from normal distribution tables.** The problem is now one of simple statistics. It is simply a matter of reading off the appropriate probability from Table 15.2 for the value of σ_T.

By linear interpolation:

$$y_v = y_1 + \left(\frac{x_v - x_1}{x_2 - x_1}\right) \times (y_2 - y_1) \qquad (15.8)$$

$$y_v = 0.758 + \left(\frac{0.715 - 0.7}{0.8 - 0.7}\right) \times (0.788 - 0.758)$$

$$y_v = \mathbf{0.76}$$

Table 15.2 Normal distribution data

σ	Probability	Cumulative probability
0.00	0.50000	0.500
0.10	0.46020	0.540
0.20	0.42070	0.579
0.30	0.38210	0.618
0.40	0.34460	0.655
0.50	0.30850	0.692
0.60	0.27430	0.726
0.70	0.24200	0.758
0.80	0.21190	0.788
0.90	0.18410	0.816
1.00	0.15870	0.841
1.10	0.13570	0.864
1.20	0.11510	0.885
1.30	0.09680	0.903
1.40	0.08080	0.919
1.50	0.06680	0.933
1.60	0.05480	0.945
1.70	0.04460	0.955
1.80	0.03590	0.964
1.90	0.02870	0.971
2.00	0.02280	0.977
2.10	0.01790	0.982
2.20	0.01390	0.986
2.30	0.01070	0.989
2.40	0.00820	0.992
2.50	0.00621	0.994
2.60	0.00466	0.995
2.70	0.00347	0.997
2.80	0.00256	0.997
2.90	0.00187	0.998
3.00	0.00135	0.999

Figure 15.9 *Linear interpolation*

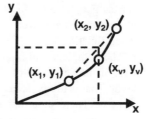

Note that it may be necessary to use linear interpolation to obtain a reasonably accurate value as shown in Figure 15.9.

It may be concluded from the analysis above that there is roughly a 0.76 probability of the project being completed by the deadline.

15.4.3 Limitations of the method

It should be noted that the above method has some limitations. First, if some of the non-critical paths are highly uncertain, they may become critical under some circumstances. Second, the β Distribution may not be an appropriate representation of the uncertainty of a given task. Third, the variation in the individual tasks may not be independent. Finally, if the number of tasks on the critical path is not large, the assumption that the aggregate distribution is normal may not be valid. Despite these limitations, however, the approach can give a useful indication of the likelihood of the project being completed with the allowable time.

15.5 Use of packages

The use of computer packages for project planning is now very common. PC-based packages (such as 'Microsoft Project') are now relatively cheap to purchase. Use of such packages certainly minimizes manual effort in project planning and greatly increases the quality of reporting. Packages can also provide multi-user access, which is important where projects involve large numbers of people spread over a wide geographical area. Packages can, however, encourage over-planning of projects and caution should be exercised in this regard. In summary, packages can be very valuable and their benefits increase with project complexity.

15.6 Project management practice

Time spent on project planning is usually rewarded by better performance. Conversely, lack of planning is usually punished. It has been seen that most projects are subject to uncertainly. Sometimes, this causes people to avoid planning. In situations of uncertainty, however, plans are perhaps even more important than cases where all tasks are deterministic.

The complexity of plans must reflect the project under consideration. One of the key skills of a project manager is to plan at the appropriate level. That is to say, breaking down the project into a reasonable number of tasks. There should be enough tasks to fully describe the project, but not so many that maintenance of the plan becomes onerous. Finally, plans are a guide to the project manager, not a constraint. Project managers need to respond to changing circumstances and not 'hide behind' the plan.

Exercises – Quantitative methods

Exercises – Probability and statistics

1 Complete the following table:

	Mutually exclusive	Collectively exhaustive
An animal drawn at random. *Warm blooded.* *Cold blooded.*		
A manufacturing student selected at random. *Studying manufacturing.* *Gender female.*		
A student selected at random. *Studying electrical engineering.* *Studying mechanical engineering.*		

2 If a die is thrown, what is the probability of the score being five or higher?

3 There are 50 people in a group of students. Of these, 25 are male and 25 female. Of the 25 males, 20 live in Manchester. Of the 25 females, 10 live in Manchester.
   (a) If a person is selected at random, what are the chances that that individual will be either a male or a Mancunian?
   (b) Two people are selected at random. What are the chances that two males were chosen?

(c) A person is selected at random and this individual is a female. Using Bayes' theorem what is the probability that this person is a Mancunian?

4 Draw a probability tree for the throw of two dice. Based on this, draw a probability distribution for the total score.

5 A person's IQ can be described by a Normal distribution with a mean of 100 and a standard deviation of 20. Assuming MENSA will accept people with an IQ of 140 or higher, roughly what percentage of the population will be eligible to join?

Exercises – Forecasting

1 Describe the nature of the forecasts that might be made over the short, medium and long term for the following:
- A manufacturer of automotive components.
- A manufacturer of consumer electronic devices.
- A university.
- A football club.

2 How would you characterize the following forecasting cases (intuition, extrapolation or prediction):
- Betting on the National Lottery?
- The weather forecast?
- Forecasting whether a football club will win on Saturday?

3 Consider the table below:

Month	Sales
Dec	1631
Jan	2504
Feb	3891
Mar	4440
Apr	5691
May	6860
Jun	7599
Jul	8678
Aug	9130
Sep	10135
Oct	11303
Nov	12215

- Calculate a first order smoothed forecast line ($\alpha = 0.3$).
- Calculate a second order smoothed forecast line ($\beta = 0.3$).

4 Sales of aftermarket disc brake pads are subject to seasonal demand as shown in the diagram below. Comment on the suitability of exponential smoothing to forecast demand of particular items.

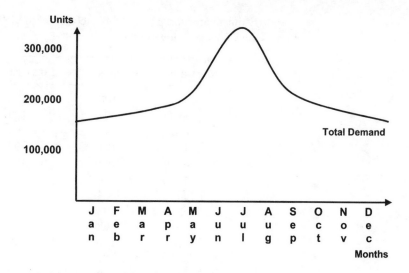

5 Calculate the line of best fit and correlation coefficient for the following data set:

X	Y
168	135
173	128
181	132
192	119
200	131
207	118
210	114
217	120
221	118
229	122
233	108
237	112
240	104
242	106
244	117
251	102
262	106
269	96
273	88
283	85
292	82
301	96
311	90
320	86
325	68
334	68
341	76
345	67

6 Estimate an appropriate safety stock level to yield a service level of 90% for the following data.

Date	Sales	Forecast
Dec	100	100
Jan	97	100
Feb	98	99
Mar	101	99
Apr	107	100
May	98	101
Jun	94	100
Jul	97	99
Aug	98	99
Sep	100	99
Oct	91	99
Nov	99	97
Dec	103	98
Jan	99	99
Feb	98	99
Mar	107	99
Apr	92	100
May	97	99
Jun	96	98
Jul	92	98
Aug	102	97
Sep	91	98
Oct	99	96
Nov	95	97
Dec	103	96

7 What is the fairest measure of a student's ability – a single module mark or the average mark for the entire year? Give reasons for your answer.

Exercises – Optimization

1 What objective functions might be employed by the following organizations:
- A university?
- A football club?
- Friends of the Earth?

2 What is the maximum value, in the range +5 to −5, of the function below:

$$f(x) = 2x - 2x^2 + 10$$

3 A hire car company purchases vehicles for £13 K per year. It is concerned with calculating the optimum number of years, x, cars

should be retained for hire before they are sold. There are a number factors that influence this decision:

- The revenue generated by the car. As the car gets older, it is off the road for longer periods. Revenue generated as a function of years retained is:

    $$\text{Revenue} = 11x^{0.8}$$

- The cost of vehicle maintenance. Again, as the car gets older, more maintenance is required. Cost is given by the following equation:

    $$\text{Cost} = 3x^{1.5}$$

- Selling price. The vehicle can be sold, but will not achieve its original price. The selling price is given by the following equation:

    $$\text{Selling price} = 11 \times 0.6^x$$

What is the optimum value of x? Also, comment on the sensitivity of the optimization to the various constants. Also, is it critical vehicles are sold exactly at the point of maximum benefit?

4 A company provides a service for moving materials from one warehouse to another. The company employs drivers and loaders, but does not own any vehicles. These are hired from another company. There is an agreement that up to 10 X type lorries or 10 Y type vans can be hired per week. The total number of drivers is 12. The loaders have limited capacity and this limitation can be described by the inequality $176 > 16X + 11Y$. For reasons of cashflow, the company has a limitation on the number of vehicles it can hire. This is described by the inequality $975 > 5X + 8Y$.

The objective function is to maximize sales revenue. This is given by the expression:

$$\text{Revenue} = 600X + 800Y$$

What are the optimum numbers of vans and lorries that should be hired and what is the revenue that can be generated? Is the business profitable in the long term?

Exercises – Evaluating alternatives

1 A company is considering building a new factory to meet possible increased market demand. The company could build either a large or small factory. If market demand increases, the resulting cashflow from building a large factory would be £40 M as compared to £20 M for building a small factory. If, however, market demand remains static, building a new factory would result in losses. In this case, there would be adverse cashflows of –£36 M and –£4 M for building a large

or small factory respectively. The probability of increased demand is 0.8. The probability of static demand is 0.2.

(a) Produce a decision table and draw a decision tree to represent the information above.
(b) Calculate the expected monetary value (EMV) for the alternatives specified in the decision table and thus select the best alternative.
(c) A management consultancy firm has approached the company and offered to undertake market research with the aim of gathering information on future demand. Their fee is £20 000. Should the company consider this offer seriously? Explain your answer.

2 A company is considering manufacturing a new action figure (Product 1) for children. There is equal probability that the response to the product will be good or bad. If the decision is made to make the action figure, then it will be possible to sell accessories (Product 2). If response to the action figure is good, then there is a 0.75 probability the response to the accessories will also be good. If, however, the response to the action figure is bad, the probability of the accessories being successful falls to 0.25. The situation can be summarized in the decision tree below.

Outline a rational decision-making process.

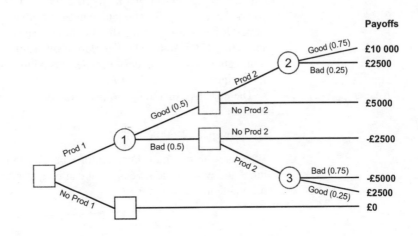

Exercises – Project management

1 A company overhauls a piece of plant as shown in the activity on arrow (AOA) network below.

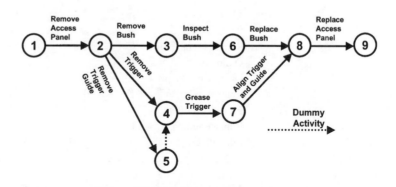

The activity times (in minutes) for the tasks are shown in the table below.

Task	Description	Time
(1–2)	Remove access panel	5
(2–3)	Remove bush	15
(2–4)	Remove trigger	25
(2–5)	Remove trigger guide	10
(3–6)	Inspect bush	5
(4–7)	Grease trigger	10
(6–8)	Replace bush	10
(7–8)	Align trigger and guide	10
(8–9)	Replace access panel	10

(a) Draw an activity on node (AON) diagram to represent the project.

(b) Calculate the overall time to complete the project. Calculate the amount of slack associated with each task.

2   A student's car needs an oil change. This will require the following
    tasks (all times in minutes):

Description	Task	Immediate predecessor	Optimistic	Most likely	Pessimistic
Warm engine	A		14	15	16
Release oil filter	B	A	1	10	19
Remove sump plug	C	A	2	15	28
Drain oil	D	C	9	10	11
Dispose of old oil	E	D	8	10	12
Remove oil filter	F	B, D	1	2	3
Fit new oil filter	G	F	1	2	3
Replace sump plug	H	F	4	5	6
Fill with oil	I	G, H	5	6	7

Making appropriate assumptions, what is the probability that the oil
change can be completed in one hour or less?

Part 4

Planning, Scheduling and Logistics

The red sticker system

January

It was Monday morning at Travis Engineering Ltd. The company was an aerospace subcontractor with 75 employees. Its speciality was high-precision machining and it supplied most of the well-known names in the aerospace industry. All was not well and the following people were meeting in the boardroom:

Harry Bardini   – Managing Director (**MD**)
John Sowden   – Works Manager (**WM**)
Michael Armitage – Production Controller (**PC**)

MD   Well guys, we've problems. We've got a load of big orders coming up. Mike, what's our delivery status?

PC   As of today, we have 34 orders overdue, 15 of them more than three weeks late. Our on-time delivery percentage is 38%, which could be better.

MD   'Could be better' is one way of putting it. 'Absolute rubbish' is another.

PC   Well, we do need to get control of priorities. . . .

WM   I thought that's why we bought that computer system to take in orders and schedule them through the factory. It certainly cost enough.

MD   Now is not the time to worry about the computer system. We need a short-term solution to the problem of getting orders out the door, even if it's quick and dirty.

WM   I've got an idea. We'll use red-stickers. We'll get our expeditors to put a red spot on any urgent job. Then we'll tell the operators to do the red-sticker jobs first. I'll crack the whip too.

MD   Sounds good. We can start today.

PC   I'm not sure about this. It all sounds too easy. Maybe we should consider pushing some of those new orders back.

MD   You are joking aren't you? We need the revenue from those orders. Not only that, I don't think our customers will accept revised delivery dates. Frankly, I don't see any way the red sticker system can go wrong. A job either has a red sticker or not, it's as simple as that.

February

MD   Well, we seem to be worse off than ever. What's our status, Mike?

PC   Our on-time percentage is now 29%.

WM   The red-sticker system has been working well though. It's really helped the operators pick the important jobs.

PC   The problem is that almost every job out there has got a red spot on it. It makes a mockery of the whole thing.

WM   Mike's right. The red-spot system has been good so far, but we need to move forward. From now on, the particularly urgent jobs should have two red stickers.

PC   Er . . . maybe we should think this through for a minute.

MD   I think John's got a point. Implement the new system immediately.

March

MD   I can't believe it. I've just walked through the shop floor and it looks like it's got measles. There's so many red spots out there its given me a headache. What's gone wrong?

PC   It wasn't long before every job had got two stickers on it, so we moved to three and before long things just sort of . . . got out of hand. All in all, it's probably a good thing we ran out of stickers.

WM   My operators still don't know what to work on. We've got to find a better way of prioritizing work.

PC   The first thing we need to do is get rid of all those damned red stickers and start from scratch.

MD   We haven't got time for that!

WM   I've got it! We'll start using blue stickers for REALLY important jobs.

Analysis

Travis Engineering has not really understood the problem that faces them. Like many real-world organizations, they are attempting to solve a fundamental problem in planning and control by simply shuffling the sequence in which they make product. If there is insufficient capacity to meet demand, sequencing cannot help. This is analogous to governments printing more money to solve an economic crisis. Any products without red stickers will become overdue and urgent in their turn. Eventually, all jobs must become late.

Summary

Planning, scheduling and logistics are key issues for manufacturing companies. This is an area that has attracted enormous interest in recent years. This part will explain the main approaches manufacturing companies use to control their factories. In addition, this part will also examine two other complementary areas: quality and maintenance management.

Objectives

- To appreciate how planning, control and logistics are an important element in the success of any manufacturing organization.
- To understand the operation of the main techniques used for planning, control and logistics.
- To understand modern concepts in quality and maintenance management.
- To be able to choose and apply different techniques in different circumstances.

16 Open-loop control systems

16.1 Overview

Factories have existed since the mid-1700s. A scientific and systematic approach to management was not developed until the twenteeth century with practitioners such as FW Taylor, Henry Gantt and Frederick and Lillian Gilbreth.

The earliest form of planning and control systems were so-called *open-loop* systems. These systems were designed to trigger the release of orders to maintain the operation of the manufacturing system. The use of open-loop control systems dates back to the 1950s and some of the techniques developed then are still in use today. There are two main open-loop approaches: reorder point (ROP) and material requirements planning (MRP) and these will be discussed in this chapter.

16.2 Basic definitions – items and BOMs

Before discussing planning and control, it is important to define some key terminology.

Items. The term 'item' refers to an entity that a manufacturing organization makes, buys or sells. An item might refer to a saleable item; in the case of an automobile company this would be a finished car. Raw materials or a bought-in finished part are also items. Finally, components or subassemblies are also components. Continuing the example of an automotive company, an engine would also be termed an item. Items are generally given a unique identifier when a computer-based control system is being used.

Bills of material (BOMs). The BOM defines the relationship between different items. Consider a complete electric motor. This is constructed from a number of components and subassemblies as illustrated in Figure 16.1.

Figure 16.1
*Subassemblies and
components of an
electric motor*

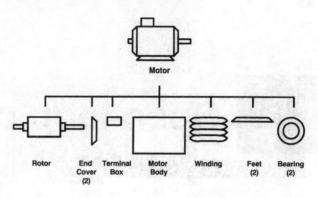

Note that to assemble a motor, a single winding is required, whereas
two end covers are needed. The rotor assembly is made from two items:
a rotor core and a shaft. The shaft itself is manufactured from 300 mm
of steel bar (see Figure 16.2).

Figure 16.2
*Further breakdown
of an electric motor*

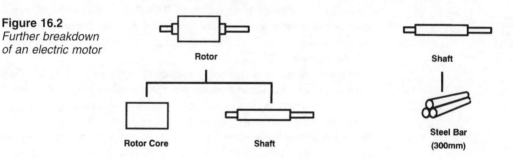

Figure 16.3 *Full
parts explosion for
an electric motor*

A complete representation of the motor would be termed a 'full part
explosion'. This is illustrated in Figure 16.3.

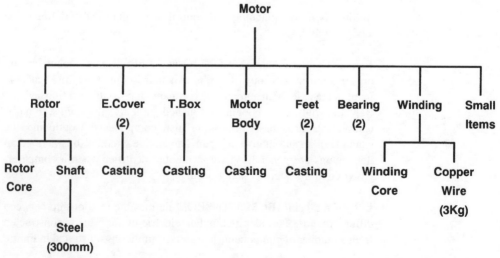

BOMs are also described as product structures. More details of how items and BOMs are stored and controlled in computer-based systems are presented later in this part.

16.3 Order models of control

Open-loop systems are concerned with order release. It is important to discuss what is meant by the term 'order'. Most manufacturing organizations base their control systems around three order types:

- **Sales.** This is raised by a manufacturing company in response to demand from a customer. It defines the following information:
  - Customer name, delivery address.
  - Product(s)/quantity(ies) required (there may be more than one item on an order).
  - Price(s).
  - Terms and conditions.
  - Delivery date(s).
- **Purchase.** This is raised by a manufacturing company in order to procure items from a supplier. It defines the following information.
  - Supplier name, delivery address.
  - Product(s)/quantity(ies) required (there may be more than one item on an order).
  - Price(s).
  - Terms and conditions.
  - Delivery date(s).

  Notice that sales and purchase orders have a similar format. Indeed, the operation of a material supply chain can be characterized by the exchange of a sales and purchase order.
- **Works.** Sometimes known as a manufacturing order. This is an internal order that authorizes the conversion of raw materials into a different form. It contains the following information:
  - Product/quantities.
  - Launch date.
  - Due date.

The interaction of these orders is shown in Figure 16.4. The stages material passes through are as follows:

1 **Purchase order receipt.** Material is received into the company. Invoice payment is authorized.
2 **Works order issue.** Material is issued to the particular works order. The material will be issued from Raw Material (RM) stores if the item on the works order is at the bottom of the BOM. Alternatively, if the works order is made from an item that has been produced in the company, material will be issued from a stock (STK) location.
3 **Shop floor feedback.** This is reported from the shop floor to monitor progress (optional).

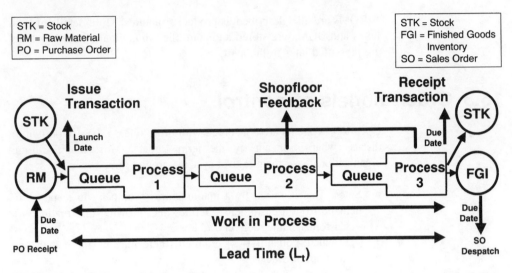

STK = Stock
RM = Raw Material
PO = Purchase Order

STK = Stock
FGI = Finished Goods
Inventory
SO = Sales Order

Figure 16.4 *Order model of control*

4 **Works order receipt.** Material is received into finished goods inventory (FGI) if it is at the top of the BOM (i.e. if it is a saleable item). If the material requires further processing and will subsequently be issued to another works order, it will be received into a stock (STK) location.

5 **Sales order despatch.** Material is despatched to customer. An invoice is generated.

This is an execution system. The question arises, however, when should orders be launched?

16.4 Basic reorder point theory

Reorder point (ROP) theory was developed in the 1950s to solve the problem of when to launch works and purchase orders. The basic theory is illustrated in Figure 16.5. The basic idea is to calculate when the stock

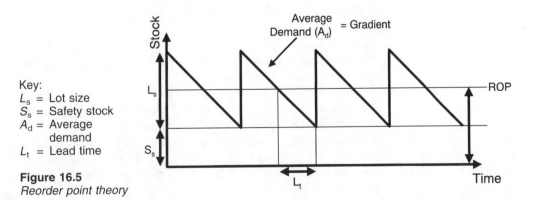

Key:
L_s = Lot size
S_s = Safety stock
A_d = Average demand
L_t = Lead time

Figure 16.5
Reorder point theory

will be consumed and launch an order so that it will arrive in time to prevent a stock-out occurring.

The basic equation is as follows:

$$\text{ROP} = A_d L_t \qquad (16.1)$$

In practice, however, it is necessary to build in a term for safety stock. This is to compensate for uncertainty in demand. The full equation is:

$$\text{ROP} = A_d L_t + S_s \qquad (16.2)$$

16.5 ROP limitations

ROP has three main limitations:

- Orders arrive late. All planning approaches are vulnerable to this effect.
- Material is consumed discontinuously.
- The value of average demand does not represent future consumption.

Consider Figure 16.6. The demand pattern has an average value of 270 units per week. Assuming a lead time of 3 months and no safety stock, this implies an ROP of 810. Simulating this situation with a lot size of 1080 (see Figure 16.7) leads to a number of *stock-outs*. This is because the demand is not uniform.

ROP theory was the first attempt to systematically control manufacturing. It has flaws, both theoretically and practically (these flaws will be discussed in Chapter 15). While these ideas are no longer applied (at least in the way that was originally intended), they are important in understanding the fundamental principles of manufacturing management.

Figure 16.6
Demand pattern

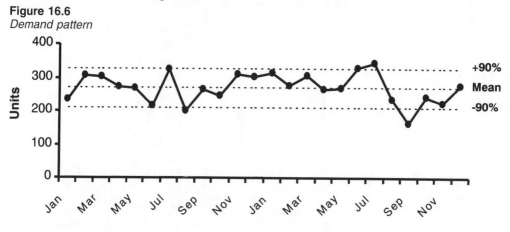

Figure 16.7   *ROP simulation*

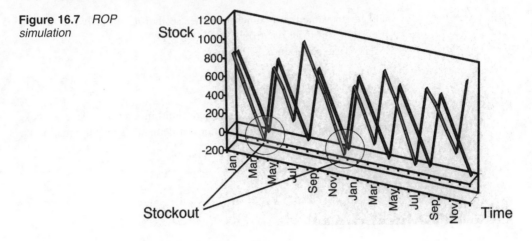

16.6 Use of safety stock

For the previously stated reasons, it is necessary to use safety stock (S_s). The basic idea behind safety stock is to buffer the planning system against uncertainty in demand. The reason for this is that it is not possible to forecast precisely. The question arises, what level of safety stock is required?

This is a highly complex issue and a full examination is beyond the scope of this book. The general principles, however, are fairly straightforward. The demand in the previous example can be plotted as shown in Figure 16.8.

The following, previously defined equations can be used to define statistical parameters associated with the demand defined above.

$$\text{Mean } (M) = \frac{1}{n} \sum_{i=1}^{i=n} x_i$$

$$\text{Standard deviation } (\sigma) = \sqrt{\frac{1}{n} \sum_{i=1}^{i=n} (x_i - M)^2}$$

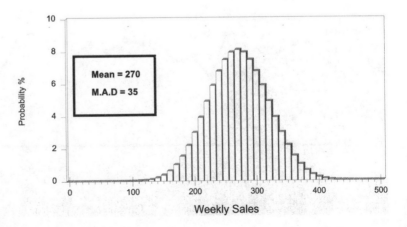

Figure 16.8
Demand distribution

$$\text{Mean absolute deviation (MAD)} = \frac{1}{n} \sum_{i=1}^{i=n} |(x_i - M)|$$

If the MAD was zero for a particular case, then there would be no requirement for safety stock. If, however, the MAD was greater than zero, then setting the safety stock to zero would lead to a service level of 50% (i.e. half of all orders would arrive late to meet demand). To improve the service level, it is necessary to build in safety stock.

The amount of safety stock required depends on the service level required. In theory, for a normal distribution the level of stock required is infinite. A safety stock of ±3σ will yield a service level of 99%. The starting point for this problem is to establish a *minimum acceptable service level*. The mathematics associated with this problem mean that as service levels tend towards 100%, the safety stock required tends towards 4. A typical compromise value is 95%. Based on this figure, it is possible to calculate an appropriate level of safety stock.

This is a three-stage process:

Stage 1. It is first necessary to calculate the MAD for the forecast as compared to actual values. Consider Figure 16.9. This shows actual

Figure 16.9
Forecast errors

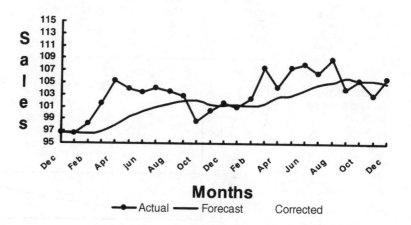

Months

━●━ Actual ━━━ Forecast   Corrected

sales and a smoothed forecast. For every data point, the absolute deviation can be calculated as shown in Table 16.1.

Stage 2. Calculate the MAD value that corresponds to the service level selected. This information is available from statistical tables and is shown in graphical form in Figure 16.10. In the example shown, the number of MADs that corresponds to a 95% service level is 1.32. Table 16.2 shows the information in Figure 16.10 in tabular form.

Stage 3. The final stage is to multiply the MAD associated with the data (in this case 2) by the number of MADs corresponding to the service level (in this case 1.32). Thus the appropriate level of safety stock is 2.64. Since the product can only be held in discrete quantities, the safety stock should be 3.

Table 16.1   Absolute deviations

Date	Actual	Forecast	Dev	Ab Dev
Dec	97	97	0	0
Jan	97	97	0	0
Feb	98	97	1	1
Mar	101	97	4	4
Apr	105	98	7	7
May	104	99	5	5
Jun	103	100	3	3
Jul	104	101	3	3
Aug	104	102	2	2
Sep	103	102	1	1
Oct	99	102	−3	3
Nov	100	101	−1	1
Dec	102	101	1	1
Jan	101	101	0	0
Feb	102	101	1	1
Mar	107	101	6	6
Apr	104	103	1	1
May	107	103	4	4
Jun	108	104	4	4
Jul	107	105	2	2
Aug	109	105	4	4
Sep	104	106	−2	2
Oct	105	105	0	0
Nov	103	105	−2	2
Dec	106	105	1	1

MAD = 2

Figure 16.10
Service level vs MAD

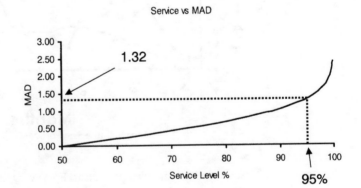

Table 16.2   Service vs MAD

Service	MAD	Service	MAD	Service	MAD	Service	MAD
50.00	0.00	78.81	0.64	94.52	1.28	99.18	1.91
53.98	0.08	81.59	0.72	95.54	1.36	99.38	1.99
57.93	0.16	84.13	0.80	96.41	1.44	99.53	2.07
61.79	0.24	86.43	0.88	97.13	1.52	99.65	2.15
65.54	0.32	88.49	0.96	97.72	1.60	99.74	2.23
69.15	0.40	90.32	1.04	98.21	1.68	99.81	2.31
72.57	0.48	91.92	1.12	98.61	1.76	99.87	2.39
75.80	0.56	93.32	1.20	98.93	1.84		

16.7 EOQ theory

Closely associated with ROP theory are principles for determining economic order quantities (EOQ). The basic idea is to balance the production benefits of large lot sizes with the stock benefits of small lot sizes. This is shown graphically in Figure 16.11. If the lot size is small, then the average stockholding is small. There is, however, a penalty because of the number of changeovers required. The converse is true if a large lot size is selected.

Figure 16.11
Effect of lot size on ordering frequency

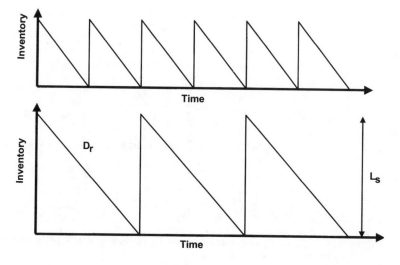

The question arises, what is the optimum compromise between production and inventory costs? Analytically, the EOQ can be determined using the formula originally devised by Wilson in the 1950s.

Annual stock holding cost = C_i (£/unit/year)
Production cost rate   = C_p (£/min)
Set-up time   = T_s (min)
Lot size   = L_s (units)
Process time   = T_p (mins)
Annual demand rate   = D_r (units/year)

The number of lots to be manufactured per year is given by:

$$\text{Number of lots} = \frac{D_r}{L_s}$$

The production cost of a single batch is as follows:

$$\text{Production cost} = C_p(L_s T_p + T_s)$$

Thus the annual cost of production is given by:

$$\text{Annual production cost} = \frac{D_r C_p (L_s T_p + T_s)}{L_s}$$

The annual inventory cost is given by:

$$\text{Annual inventory cost} = \frac{1}{2} L_s C_i$$

Thus the total cost is given by:

$$\text{Annual total cost} = \frac{D_r C_p (L_s T_p + T_s)}{L_s} + \frac{1}{2} L_s C_i$$

To determine the optimum lot size, differentiate with respect to L_s:

$$\frac{dD_c}{dL_s} = \frac{1}{2} C_i - \frac{D_r C_p T_s}{L_s^2}$$

Setting to zero and rearranging for L_s to give the optimum lot size, L_{sopt}.

$$L_{sopt} = \sqrt{\left(\frac{2D_r C_p T_s}{C_i} \right)} \qquad (16.3)$$

This equation can be displayed graphically as shown in Figure 16.12.

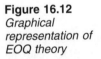

Figure 16.12
Graphical representation of EOQ theory

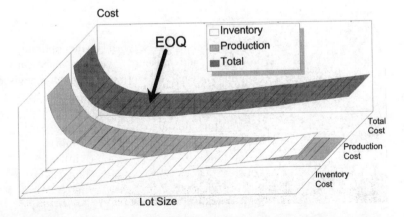

16.8 Summary

ROP/EOQ theories were the first attempts to systematically control manufacturing. They have flaws, both theoretically and practically. While these ideas are no longer applied (at least in the way that was originally intended), they are important in understanding the fundamental principles of manufacturing management.

16.9 Material requirements planning (MRP)

In most manufacturing companies, ROP has serious limitations when applied alone. The development of MRP offered a solution. It was also the first stage in the evolution of manufacturing resource planning (MRPII), which was widely used in the 1970s and 1980s. According to the American Society for Production and Inventory Control (APICS), MRP is defined as follows:

> *The planning of components based on the demand for higher-level assemblies.*

This rest of this section will describe the mechanics and advantages of this approach.

16.9.1 Dependent and independent demand

In the previous chapter, it was shown that reorder points could be calculated for a particular item to establish when new orders should be launched. It was further demonstrated that safety stock was required to compensate for uncertainty in demand. To calculate the appropriate level of safety stock, it is necessary to define an acceptable level of *service*. This principle is fundamentally flawed for an assembled product. Imagine an assembly with ten components as shown in Figure 16.13.

Figure 16.13
Example of a BOM

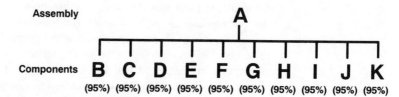

If the service level for each individual component is 95%, it might be thought that the service level for the assembly is also 95%. This is erroneous: to manufacture assembly A, *all* of the components must be available. The probability of having all of the components is $(0.95)^{10}$, which represents a service level of less than 60%. Even if sufficient stock to give an individual service level of 99% is provided, the service level for the whole assembly would be around only 90%. The reason for this is that there are two demand types: independent and dependent.

The basic principle is that items with dependent demand should *not* be planned using ROP. The exception to this rule is if the item has the following characteristics:

● **Cost.** The item represents a small fraction of the total value of the assembly.

- **Volume.** The item is used in large quantities.
- **Commonality.** The item is used on several different assemblies.
- **Non-critical.** Alternative items can be used.
- **Availability.** The item is available from a variety of sources at short notice.

Such items are often classed as 'C' type. In most manufacturing companies, a relatively small number of items have a disproportionate influence on the total cost of a product. This is an example of *Pareto's law* (sometimes also called the 80–20 rule). Pareto analysis involves plotting the cumulative value of a parameter for a population against the proportion of the same population having that cumulative value. This can be illustrated in Figure 16.14.

Figure 16.14
Pareto curve for components in an assembly

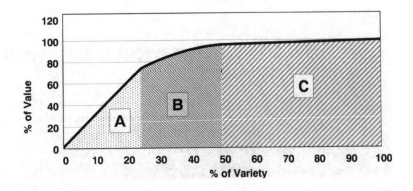

Pareto analysis is valuable in a variety of business situations. For example, in most manufacturing organizations, 20% of customers will be responsible for 80% of sales (in rough terms). For parts in an assembly, it is likely that a relatively small proportion of the parts in an assembly will constitute a large proportion of its total value (as illustrated in Figure 16.14). For an internal combustion engine, the cylinder block would be classed as an 'A' class item. A more moderately priced item such as the cylinder head gasket would probably be classed as a 'B' value item. Fasteners, however, would undoubtedly be classed as 'C' items. For such low value items, it is feasible to hold large safety stocks and thus a simple approach such as ROP is appropriate.

16.9.2 MRP mechanics

MRP operates differently from ROP. Instead of considering each item independently, MRP considers the relationships between different items. MRP starts from sales demand and first converts this *gross requirement* into a *net requirement*. In order to accomplish this, the following data need to be considered:

- **Inventory status.** Requirements are not created if there is enough stock-on-hand (SOH).

- **Order status.** Again, requirements are not created if there are enough orders pending.
- **Item data.** This includes lead times, lot sizes and any allowances for scrap or yield.

The basic formula to determine net requirements is as follows:

$$P_{ai} = P_{a(i-1)} + O_{si} - R_{gi} \qquad (16.4)$$

Where i = period for calculation, P_{ai} = projected available balance, O_{si} = scheduled receipts and R_{gi} = gross requirements.

If the projected available balance is less than zero:

$$R_{ni} = P_{ai} \qquad (16.5)$$

where R_{ni} = net requirements.

Net requirements need to be adjusted to meet conditions outlined in the item data to create *planned orders*. The launch date for a planned order is calculated by subtracting the lead time (from item data) from the date required. Finally, the requirements can be projected into the components of the item. The process will then continue at all levels in the *bill of materials* (BOM).

A0

|

F1

Figure 16.15
Example bill of material

16.9.3 MRP calculation

The MRP calculation can be illustrated by the following example. Consider a simple product, A0, that is made from a single component, F1 (see BOM in Figure 16.15).

Part number: <u>A0</u>   Lead time: <u>3</u>   Lot size: <u>60</u>

	SOH	1	2	3	4	5	6	7	8
Proj. gross requirements		20	20	20	20	20	20	20	20
Scheduled receipts		60							
Proj. available balance	25	5	45	25	5	45	25	5	45
Planned order receipt						60			60
Planned order release			60			60			

Note that the figures running along the top of the table, 1 to 8, represent week numbers. The starting point for the calculation are the figures shown in bold. Because A0 is at the top of the BOM, the projected available balance represents customer orders.

Using equation (16.4), the projected available balance for week 1 is determined by subtracting the projected gross requirements for week 1 from the SOH. Thus:

Week 1: $5 = 25 - 20$

The projected available balance for week 2 is calculated in the same way, except on this occasion the arrival of material due to the scheduled receipt in week 2 must be taken into account:

Week 2: $45 = 5 - 20 + 60$

The calculation for weeks 3 and 4 is similar to that for week 1:

Week 3: $25 = 45 - 20$
Week 4: $5 = 25 - 20$

In week 5, if the calculation was simply repeated, the projected available balance would yield a negative value ($5 - 20 = -15$), which has no physical meaning. The effect of this is to prompt the creation of a planned order. The planned order would need to be received in week 5. Note that because the lot size is set at 60, it is this quantity that is used in the planned order receipt row:

Week 5: $45 = 5 - 20 + 60$

MRP calculates the start date for the planned order by subtracting the lead time (3 weeks) from the due date. Thus the planned order release in week 2 corresponds with the planned order receipt in week 5. The calculation of projected available balance for weeks 6 and 7 is straightforward:

Week 6: $25 = 45 - 20$
Week 7: $5 = 25 - 20$

In week 8, however, the projected available balance would again become negative unless another planned order is created. The same logic used for week 5 is repeated. Thus:

Week 8: $45 = 5 - 20 + 60$

Note that because neither of the planned orders is due for release in the current period, MRP would take no action for A0 during the current week.

The projected gross requirements for F1 are derived from the planned order releases for A0. This can be best understood by considering the physical manufacture of A0. For the manufacture of the 60 units of A0 to commence in week 2, 60 units of F1 will need to be available at this time.

The calculations applied for A0 can be repeated for F1. Thus:

Week 1: $15 = 15 - 0$

In week 2, a planned order will need to be created:

Week 2: $75 = 15 - 60 + 120$

The calculations for weeks 3–8 are straightforward. As with A0, MRP will take no action for F1 as no planned orders are due for release. If the calculation is repeated in a week's time, however, and there are no other changes, the release date for the planned order will move into the current week. This will cause an order to be generated.

Part number: <u>F1</u>   Lead time: <u>1</u>   Lot size: <u>120</u>

	SOH	1	2	3	4	5	6	7	8
Proj. gross requirements			60		60				
Scheduled receipts									
Proj. available balance	15	15	75	75	75	15	15	15	15
Planned order receipt			120						
Planned order release		120							

Note that the relatively stable demand of A0 results in a single order for 120 units of F1. This is *not* a function of the MRP planning system, but a consequence of the lot sizes applied.

16.10 Summary

MRP is more complex than ROP and also requires a bill of material. It does, however, overcome the main limitations of ROP (ROP is still appropriate for C class items). It will be seen in later chapters that MRP is the scheduling engine in closed-loop MRP and MRPII.

17 Closed-loop MRP

17.1 Overview

MRP overcomes the obvious limitations of ROP. Simple MRP is sometimes referred to as open-loop MRP or MRPI. Its overall objectives and philosophy, however, are similar to ROP: that is to say, it is concerned with *launching orders. Closed-loop MRP* goes further in that it not only launches orders, but monitors the actual activity of the system.

Closed-loop MRP was heavily promoted by the American Production and Inventory Control Society (APICS) in the 1970s. This promotion was so vigorous that it later became known as the MRP 'crusade'. It is defined as follows:

> *A system built round MRP, but including the additional functions of sales and operations planning, master production scheduling and capacity requirements planning. Closed-loop implies monitoring and control of the execution functions (both purchase and manufacture).*

This chapter will give an overview of closed-loop MRP. The additional elements (e.g. master production scheduling) not discussed so far in the book will be reviewed later. This chapter will concentrate on the use of MRP itself within the context of the complete closed-loop structure.

17.2 Closed-loop control

The concept of open- and closed-loop systems is derived from control theory. Closed-loop MRP operates in a cycle as shown below:

- **Planning.** MRP determines when orders need to be launched (see logic in chapter 16).
- **Execution.** Orders are released into manufacturing or to suppliers.

- **Feedback.** Changes in conditions are reported to the system.
- **Corrective action.** The system recommends corrective action.

This approach can be contrasted with MRPI where orders are simply *launched*. This is based on the assumption that the environment will remain static. In practice, conditions are changing constantly for a number of reasons. These include:

Sales variations. Customers may change their requirements, possibly at short notice. They may also change quantities required.

Supplier variations. Suppliers may fail to deliver product on time. They may fail to deliver the correct quantity, or some of the products supplied may be unusable due to quality problems.

Manufacturing variations. The factory may deliver orders late or possibly scrap some items, leading to a reduced quantity made.

Stores problems. Stock values held on the computer system may not reflect the actual quantity present. Alternatively, some items in stores may be of poor quality.

Engineering changes. Design engineering personnel may change the BOM of a product or redefine the method by which a product is manufactured.

Planning changes. Planning policies such as lead times or lot sizes may be changed.

If conditions change, even slightly, this can have a considerable effect. Consider the following.

Part number: A0   Lead time: 3   Lot size: 60

	SOH	1	2	3	4	5	6	7	8
Proj. gross requirements		40	20	20	20	20	20	20	20
Scheduled receipts			60						
Proj. available balance	25	45	85	65	45	25	5	45	25
Planned order receipt		60						60	
Planned order release	60				60				

Part number: F1   **Lead time: 1**   **Lot size: 120**

	SOH	1	2	3	4	5	6	7	8
Proj. gross requirements	60				60				
Scheduled receipts									
Proj. available balance	15	75	75	75	15	15	15	15	15
Planned order receipt	120								
Planned order release	120								

Compare this example to that in the previous chapter. In this case, the only change is that the gross requirements for item A0 in week 1 have been increased from 20 to 40. This has an impact not only on item A0, but also F1. In the case of A0, the system requires an order to be launched immediately and completed in only one week (note that the standard lead time is three weeks). This order would need to 'leapfrog' the scheduled receipt for 60 units due in week two. In the case of F1, 120 units need to be launched and produced immediately. Clearly, the increased demand for A0 in week one cannot simply be handled by launching orders. Thus, open-loop MRP is inadequate for this case.

In practice, when conditions change in the manufacturing system so that particular items are unavailable, they will be expedited. This leads to the development of informal systems where the real mechanism of control is via a shortage list. Informal systems will ultimately undermine formal control mechanisms.

17.3 Mechanism for closed-loop control

17.3.1 Overview

Closed-loop MRP attempts to overcome the limitations of simple MRP in three ways:

- By the provision of capacity planning tools to ensure that the orders generated by MRP are feasible.
- By the monitoring of the execution of plans.
- By the printing of planner action messages.

These points will be discussed in detail. Closed-loop MRP can be represented as shown in Figure 17.1. Note that simple open-loop MRP is a single module within the closed-loop MRP system. Note also the feedback loops designed to ensure consistency between planning levels.

Figure 17.1
Closed-loop MRP

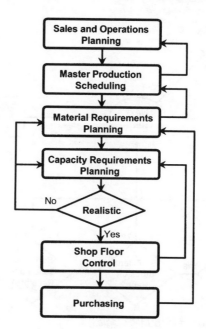

17.3.2 Capacity management

Sales and operations planning (SOP). Sometimes this is referred to as aggregate planning. This is a manual management function. It requires management to review business requirements over the medium term and provide capacity.

Master production scheduling (MPS). MPS relates to the top planning of saleable items. It allows any discontinuities between supply and demand to be taken into account. If it is not possible for demand to be met, MPS is the mechanism whereby management can exercise control.

Capacity requirements planning (CRP). CRP checks the output of MRP to ensure that sufficient capacity is available at a detailed level to meet the plans generated.

17.3.3 Monitoring of plans

Open-loop MRP simply prints orders. There is no provision to check whether these orders are proceeding to plan. Closed-loop MRP, however, receives feedback on the execution of the works orders generated via *shop floor control* (SFC). This feedback is in two forms: first, the progress of orders through the various operations required. Second, if any items in a batch are scrapped, this will similarly be recorded. Changes in purchase order due dates can also be recorded via the *purchasing* module.

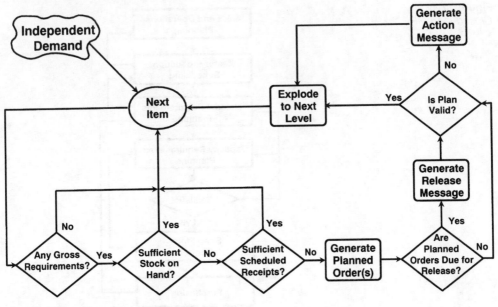

Figure 17.2 *Closed-loop MRP logic*

17.3.4 Planner action lists

The effect of changes/problems relating to plans is communicated in *planner action lists* in a closed-loop MRP system. These are generated following an MRP computer run. The logic of this process is illustrated in Figure 17.2.

It should be noted that the MRP calculation needs to be undertaken in a logical order. The complete set of items is processed from the top of the bill of material downwards. This is to ensure that *all* demands from higher level assemblies are passed to lower-level items. The resulting action lists are normally distributed to *planners* who act accordingly. Examples of messages are as shown below. Note these can apply to manufactured or purchased items.

- **Past due order.** A scheduled receipt has not been completed by the specified due date.
- **Planned order due for release.** A planned order needs to be released now.
- **Reschedule order in.** Make the completion date on an order earlier.
- **Planned order past due for release.** A planned orders release date is sometime in the past. The order will therefore need to be manufactured/purchased in less than the standard lead time.
- **Reschedule order out.** Make the completion date on an order later.
- **Cancel order.** Delete an order from the system.

Planners do not simply follow the action messages, but use these to support decision making.

17.4 Manufacturing resource planning

Manufacturing resource planning (MRPII) is similar to closed-loop MRP. Figure 17.3 illustrates the different elements of MRPII. It is defined as follows:

> *A method for planning of all resources in a manufacturing organization. It includes a simulation capability and is integrated with financial planning. These systems work with the modules of the closed-loop MRP.*

The actual mechanics of closed-loop MRP and MRPII are the same, but MRPII also introduces a simulation capability. MRPII includes a financial management facility and is a methodology for the management of the entire organization.

Figure 17.3   *MRPII*

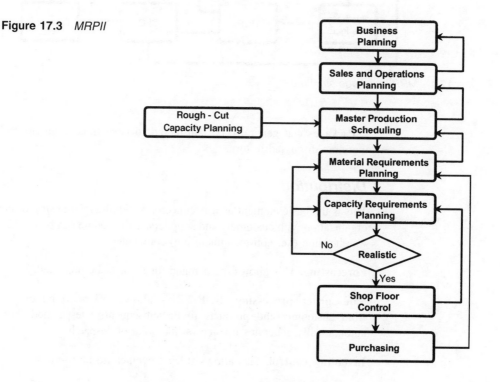

17.4.1 MRPII packages

It should be emphasized that MRPII is a generic philosophy. There were, however, a number of software packages developed for its support. These packages would be organized into a series of modules. These allowed different companies to purchase and implement the range of functions most appropriate to their organization. A typical MRPII package structure is as shown in Figure 17.4.

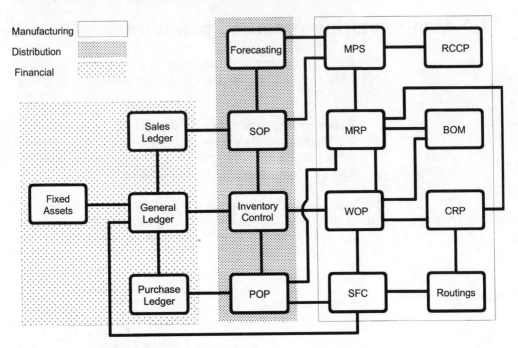

Figure 17.4 *MRPII package structure*

Packages can generally be considered to have three elements and these are discussed in turn.

Distribution

This is the basic element as it is concerned with the relationship of the organization with customers and suppliers. This module can be used by warehousing (i.e. non-manufacturing) companies.

Forecasting. This allows the demand for items to be predicted.

Sales order processing (SOP). This allows orders to be entered from customers and products to be subsequently despatched. SOP automatically generates invoices at the point of despatch.

Inventory control. This allows stock movement to be controlled.

Purchase order processing (POP). This allows purchase orders to be raised on suppliers.

Manufacturing

This is concerned with the control of the manufacturing system:

Master production scheduling (MPS). This allows the top-level plan for manufacturing to be defined. Note the interaction with the forecasting and SOP modules to support order promising.

Rough-cut capacity planning (RCCP). This allows the rough validation of manufacturing plans.

Material requirements planning (MRP). This creates planned orders and generates planner action messages based on the MPS.

Works order processing (WOP). This allows works orders (and paperwork) to be generated.

Shop floor control (SFC). This allows the progress of work orders to be monitored and detailed work-to-lists to be generated.

Capacity requirements planning (CRP). This produces a detailed capacity plan to be generated based on MRP planned orders and works orders currently in production.

Bill of materials (BOM). This allows the relationship between different items to be defined.

Routings. This defines the manufacturing sequence for a particular item.

Financial

These modules are used specifically by accounting departments. They employ information generated by transactions created within other modules. These are:

Sales ledger (also called accounts receivable). This interacts with SOP. It manages payments outstanding from customers.

Purchase ledger (also called accounts payable). This interacts with POP. It manages payments owed to suppliers and authorizes payment of invoices.

General ledger (also called nominal ledger). This is the main tool for generating accounting information for the organization.

Fixed assets. This maintains records of the fixed assets (e.g. plant) owned by the organization.

17.4.2 MRPII data structures

Underpinning MRPII packages is a standard data structure. These data are in two forms:

- **Static.** These are data that are not altered routinely via transactions. Despite the fact that these data are classified as 'static', in certain environments, changes can be frequent – for example, design

Figure 17.5 *Static and dynamic data*

Function	Static	Dynamic
• Sales/Marketing	Customer Addresses	Sales Orders
• Design	Part Geometry	
• Process Planning	Machines	
• Production Planning	Work Centre Capacity	Works Orders
• Purchasing	Suppliers Addresses	Purchase Orders
• Personnel	Personnel Names	Payments Year to Date
• Accounts	Account Numbers	Account Balances

changes in the aerospace industry can present a major workload to the companies concerned.

- **Dynamic.** These are data that are routinely updated via predefined transactions.

Examples are shown in Figure 17.5.

17.4.3 Master files

In a commercially available MRPII package, there are hundreds of files within the integrated databases with many thousands of individual fields. The main data files on which systems are built are shown in Figure 17.6.

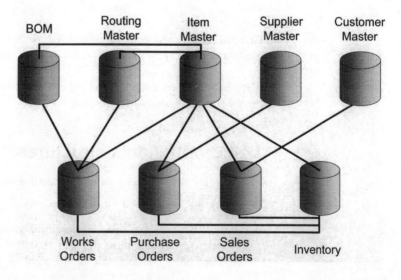

Figure 17.6 *Main master files*

The top row represents the main static data files. The bottom row represents the main dynamic data files. The static order files are as follows:

- **Customer master.** This defines information associated with a particular customer (name, address, etc.).
- **Supplier master.** This defines information associated with a particular supplier (name, address, etc.).
- **Item master.** This is perhaps the key file in the entire system. It defines each item that is made, sold or purchased by the organization. Information stored on this file includes descriptions, the batch quantity for manufacture and the lead time for supply.
- **Routing master.** This defines the process by which a particular component is manufactured. It defines the equipment needed and time taken at every operation. This file also provides textual information specifying exactly how an item will be produced.
- **Bill of material (BOM).** The BOM defines the relationship between different items. That is to say, it defines which items are required to assemble another.

The dynamic data files define the ongoing operation of the organization. The inventory file stores details of items held in the organization. The operational status of the organization is defined by orders.

17.5 Development of ERP from MRPII

The structures defined for MRPII form the core of enterprise resource planning (ERP) systems. Because the concept of ERP is relatively new, it has no generally accepted definition. Indeed, some software packages have simply been rebadged from 'MRPII' to 'ERP'. Having said this, several authors have written on the subject of ERP and there is a growing consensus on its attributes. These generally relate to the ability of ERP to manage multi-site organizations as opposed to MRPII, which is oriented towards single-site operation. The following features are commonly cited:

- The ability to create product families to allow the definition of medium-term plans at an aggregate level.
- Reporting systems to collate information from several sites.
- Tools for linking the ERP database to systems from other companies. These are sometimes referred to as *supply chain tools*. Systems for Electronic Data Interchange (EDI) are provided to allow the ERP systems of different businesses to interact directly. Tools are also provided to support schedules rather than discrete orders. This allows businesses in the supply chain to harmonize more closely their manufacturing activities. Many ERP packages are so-called 'internet-enabled' systems as they allow information to be viewed via a standard web browser.

In addition to the above, most ERP packages now offer additional modules to provide a more complete business management solution. These additional modules include the following:

Configured products. This module allows make-to-order products to be defined for particular sales orders without the need to define BOMs.

Human resource management (HRM). Functionality is provided for personnel records, time and attendance monitoring and payroll.

Maintenance management. This supports maintenance activities within the organization. These activities include spares control, condition monitoring and preventive and corrective maintenance.

Quality management. This includes sampling and statistical process control (SPC).

ERP systems have also evolved substantially in their use of IT. All ERP systems are based on integrated databases, which make the development of customized reports far easier than in previous systems. Integration with other organizational subsystems is also far simpler. Modern ERP systems are also highly flexible, allowing them to be configured to the requirements of particular business without the need for bespoke programming. This ability to use a standard package has a number of advantages:

Maintenance fees. Most companies pay a maintenance fee (typically around 15% of the price of the package) to the package vendor for software updates (if a company has substantially modified their package, it can be impossible to accept these updates in practice).

Business focus. The company can focus on business rather than software issues. This supports the business process re-engineering (BPR) philosophy. In BPR, improvement is derived from changes to business processes rather than simple automation of existing operating practices. Traditional systems could often be a barrier to change because of their relative inflexibility.

Implementation times. Implementation times can be reduced enormously.

System reliability. Software reliability can be greatly improved.

Finally, modern ERP systems take advantage of improvements in IT. For example, virtually all ERP systems are based on open systems and can be implemented on a variety of platforms. Furthermore, most ERP packages can be run on a client–server architecture, which confers a number of benefits. In particular, client–server architectures allow a graphical user interface (GUI) to be employed on workstations while still allowing access to a central database.

17.6 Summary

Closed-loop MRP is designed to overcome the main limitations of simple MRPI. It is not simply an order launching system, but employs planner action lists to allow a response to changes in the environment. In order to do this, closed-loop MRP also introduces a number of other elements such as shop floor control and master production scheduling. The principal aim of closed-loop MRP is to generate viable plans and then monitor execution.

From relatively simple origins, ERP has evolved into a sophisticated, integrated business management system. It is interesting to note that ERP is expanding beyond its manufacturing base into other sectors. For example, 'SAP' and 'Baan' have offerings for the healthcare industry. 'Baan' even has a specific offering for higher education.

It must be recognized, however, that ERP, like any information system, has two elements. First, it is comprised of software that can (in these days of highly configurable systems) be purchased as standard. What cannot be bought off-the-shelf are the business processes themselves. These depend on far more than software; they are critically dependent on human factors and effective process design. It should also be recognized that it is not a trivial task to capture business requirements and configure an ERP system accordingly. Finally, ERP systems can present a danger; they can easily cause an organization to focus on systems as a solution. In practice, simple, manual processes (Kanban, for example) have a role to play alongside computer systems.

One of the most crucial parts of closed-loop MRP is MPS. This will be discussed in detail in the next chapter.

18 Master production scheduling

18.1 Overview

Perhaps the most important module within MRPII is master production scheduling (MPS). MRPI is driven purely by customers. The purpose of MPS within MRPII is to decouple manufacturing from sales. This is counterintuitive: to optimize company performance, products would be manufactured solely to meet customer requirements. In practice, however, there are many constraints that dictate that sales orders cannot simply be used to drive manufacturing.

MPS is defined as follows:

> The anticipated build schedule for those items assigned to the master scheduler. The master scheduler maintains this schedule and in turn it becomes a set of planning numbers which drives MRP. It represents what the company plans to produce expressed in specific configurations, quantities and dates. The MPS is not a sales forecast, which represents a statement of demand. The MPS must take account of the forecast, the production plan (SOP), and other important considerations such as order backlog, availability of material, availability of capacity, management policy and goals, etc.

MPS has two functions:

- MPS is the mechanism for translating the sales and operations plans into manufacturing plans.
- MPS is the mechanism for balancing supply and demand.

This second point cannot be overstressed: one of the most common misunderstandings is that the MPS is a forecast, which is a statement of demand. MPS is a statement of supply. The reason for this is illustrated in Figure 18.1.

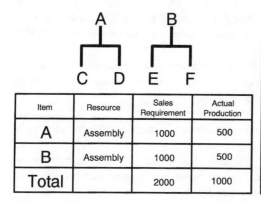

Item	Resource	Sales Requirement	Actual Production
A	Assembly	1000	500
B	Assembly	1000	500
Total		2000	1000

Assembly Shop Capacity = 2000
Machine Shop Capacity = 3000

Item	Resource	Net Requirement	Actual Production
C	Machine Shop	1000	1000
D	Machine Shop	1000	500
E	Machine Shop	1000	1000
F	Machine Shop	1000	500
Total		4000	3000

Figure 18.1 *MPS example 1 – effect of limited capacity on assembled items*

In Figure 18.1, It can be seen that there is a total sales demand for products A and B of 2000 in a given period. There is sufficient capacity to assemble A from C and D and B from E and F. The demand for A and B translates (via MRP) to a total machining requirement of 4000 units (assuming zero stock and scheduled receipt). But there is capacity to machine only 3000. Intuitively, this would suggest that in the period under consideration, 1500 of item A and 1500 of item B could be manufactured. If, however, C, D, E and F are not manufactured in matched sets, then assembly will be starved of parts. If for example, the machine shop chose to make C, D, E and F in quantities of 1000, 500, 1000 and 500 respectively, the outcome in assembly would be actual production levels for A and B of only 500 each.

In the second example, shown in Figure 18.2, there is a sales requirement for 400 000 items in a specified period. Using the bill of material (x is made of y), there is also a requirement for 400 000 of item y. If there is capacity to manufacture only 300 000 item x, this is the quantity

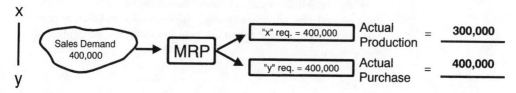

Figure 18.2 *MPS example 2 – effect of limited capacity on purchased items*

that will be produced. The quantity of y purchased, however, will be 400 000. This means 100 000 units of item y will be left in stock.

These problems occur because the requirements driving MRP are not feasible. MPS attempts to generate a feasible schedule at top level that can form the basis of lower level plans.

18.2 Reasons to master schedule

There are several reasons why sales demand cannot be used to drive MRP:

- **Lumpy demand.** If demand exceeds capacity during some periods, it will be necessary to move requirements into earlier, underloaded periods.
- **Seasonal demand.** If demand is seasonal, then it will be necessary to manufacture product for stock during those periods when capacity exceeds requirements.
- **Month-end biased.** It is common in many industries for demand to be biased to a month-end. This again means products need to be manufactured earlier than sales requirement.
- **Overload.** If the company has more demand than available capacity, it is essential that decisions are taken on which sales orders will *not* be supplied before MRP is run.
- **Set-up dependence.** Some companies need to manufacture in a particular sequence to minimize changeover disruption.

All of the above approaches will lead to excess stock being made. It should also be noted that MPS is used to damp out fluctuations in demand. Thus MPS is a time-phased manufacturing plan. A typical format is shown in Figure 18.3.

Figure 18.3
Tabular MPS format

Part No.	11	12	13	14	15	16	17	18	19	20
DP1029	1,000	1,000	1,000	1,000	1,000	1,000	1,000	1,000	1,000	1,000
DP1030	1,500	1,500	1,500	1,500	1,500	1,500	1,500	1,500	1,500	1,500
DP1031	2,000	2,000	2,000	2,000	2,000	2,000	2,000	2,000	2,000	2,000
DP1032		200		200		200		200		200
DP1033	1,000	1,000	1,000	1,000	1,000	2,000	2,000	2,000	2,000	2,000
DP1034					7,000					

Planning Horizon — Weeks

While the MPS process is based on demand (actual or forecast), it must also take into account product characteristics and factory capacity. MPS is thus the means by which management exercise control of the planning process.

18.3 MPS stability

It is crucial that the master schedule is kept relatively stable. The period over which the MPS operates is called the planning horizon. This determines how far forward in time planning calculations will be made. Changes in the MPS close to the present have greatest significance. If there is a change in the MPS within the manufacturing lead time, this will cause changes to released works orders. Such changes may be extremely disruptive. For this reason, many MRPII packages allow a demand time fence (DTF) to be defined. This warns planners if new demand is scheduled within this period: if this is the case, a time fence violation message is issued. It should be emphasized, however, that the system does not prevent orders being placed with the DTF by the planner. MPS is designed to support the planners, not replace them. Generally (though not always), the DTF will be set at the cumulative manufacturing lead time for the product.

Some MRPII packages also allow a planning time fence (PTF) to be defined, normally set at the cumulative purchasing and manufacturing lead time. In this region, works orders would normally be in the planned or firm condition. Thus changes will not be disruptive to production. It would be expected, however, that purchase orders would be in the released condition. Therefore, manufacturing trade-offs are possible, provided no additional purchase requirements are generated. In this period, MPS considers both forecast and actual demand, using the greater for calculation of projected available balance (PAB).

One issue many companies need to address is the need to service emergency demand. That is to say, a product that is required by customers within the normal manufacturing lead time. It is often possible for a company to charge a premium price for such business. In practice, some orders can be manufactured in less than the normal lead time simply by advancing these products to the front of queues waiting to be processed. This inevitably leads to other orders being given lower priority. In practice, unless the number of expedited orders is small, the effect on the manufacturing system as a whole is extremely disruptive. There are several approaches for handling emergency demand:

- **Short order shop.** Some organizations set up small, independent production facilities specifically designed to cope with emergency demand. These facilities are optimized to make small product quantities with short lead time at the expense of production efficiency.
- **Uncommitted capacity.** If there is spare capacity in the organization and some items are expedited, there will be sufficient time to 'catch up' on other orders. This approach requires great discipline in not utilizing the full available capacity.
- **Trade-offs.** If an order can be deferred (or 'de-expedited'), then it is possible to accept new demand.
- **Uncommitted inventory.** 'Spare' finished product inventory can be used to meet immediate requirements thus allowing items in current production to be expedited. Again, this approach requires discipline.

18.4 MPS mechanics

Most MRPII systems have systems to address master scheduling. While all of these systems operate differently, the overall approach is the same for all. The system compares demand in the form of actual demand and forecasts to the MPS. In most cases, the MPS is created manually as a series of time-phased firm planned orders. MRP planner action messages are printed to support the planner (for example, if the PAB

Figure 18.4
*Typical demand
pattern*

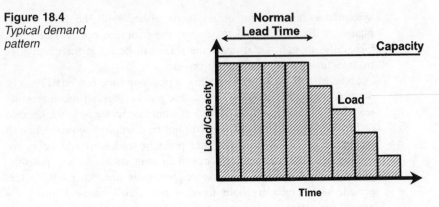

becomes negative). The system will suggest orders to be raised, but these recommendations may be over-ridden by the planner. Special messages will be generated if these order recommendations conflict with the DTF.

One tool normally provided by the system is available-to-promise (ATP) logic. In most companies, the demand profile is similar to that shown in Figure 18.4. The company will aim to fill capacity to the normal level of utilization over the period up to the normal lead time offered to customers. Beyond this period, it would be expected that firm demand would be insufficient to fill the capacity. This difference between load and capacity is generally filled by forecast. ATP logic calculates how much product can be promised without affecting existing sales orders. ATP works by comparing the firm demand with supply.

ATP is calculated in the following example. Consider the MPS grid below for weeks 1–8. This shows the forecasted sales as predicted by the sales and marketing department (*Independent demand*). The row '*Actual demand*' shows orders received to date by customers. *Projected available balance* (PAB) is calculated in the same way as the MRP example in 16.9.3 with one exception; in the case of MPS, the PAB may be negative. No planned orders are created in MPS and the PAB is simply used to create action messages. MPS orders are created manually. The '*MPS (receipt)*' row shows when MPS orders are planned to be received. Note the effect of the DTF after week 2. In this case, the DTF corresponds to the cumulative manufacturing lead time for the item. There is no forecast demand before the DTF; if actual orders have not arrived in line with the forecast by this point, there will be insufficient time to make them. After the DTF, the greater of forecast and actual demand is used for calculation of PAB. Thus in week 3, the PAB is calculated as shown below:

Week 3: $45 = 5 - 20 + 60$

In this simple example, PTF has no effect and is displayed only for the information of the user of the grid.

Part number: K7   Lead time: 3   Lot size: 60

	SOH	1	2	3	4	5	6	7	8
				DTF		PTF			
Independent forecast				20	20	20	20	20	20
Actual demand		15	5	10	20				
Proj. available balance	25	10	5	45	25	65	45	25	5
MPS (receipt)				60		60			
Available to promise									

The final row is ATP. Normally, the MPS grid would be implemented on a computer system. Customers would contact the company and ask when products are available and in what quantities. The calculations for ATP are as follows:

1 In week 1, ATP is the quantity that can be promised immediately. The quantity that can be promised will change when MPS orders are due. Thus ATP in week 1 must consider the period before the first MPS receipt in week 3 (i.e. weeks 1 and 2).

$$\begin{aligned}
\text{Total supply} &= \text{Stock + Receipts} \\
25 &= 25 + 0 \\
\text{Total demand} &= \text{Firm orders (weeks 1 and 2)} \\
20 &= 15 + 5 \\
\text{ATP}_1 &= \text{Total supply} - \text{Total demand} \\
5 &= 25 - 20
\end{aligned}$$

2 ATP will remain constant until the next MPS receipt in week 5. Thus ATP in week 3 will consider the period week 1 to week 4. Thus:

$$\begin{aligned}
\text{Total supply} &= \text{Stock + Receipts} \\
85 &= 25 + 60 \\
\text{Total demand} &= \text{Actual (weeks 1--4)} \\
50 &= 15 + 5 + 10 + 20 \\
\text{ATP}_3 &= \text{Total Supply} - \text{Total demand} - \text{ATP}_1 \\
30 &= 85 - 50 - 5
\end{aligned}$$

3 There are no further MPS receipts over the planning horizon. Therefore, ATP will need to consider *all* supply and demand. ATP in week 5 will consider the entire period, thus:

$$\begin{aligned}
\text{Total supply} &= \text{Stock + Receipts} \\
145 &= 25 + 60 + 60 \\
\text{Total demand} &= \text{Actual (weeks 1--8)} \\
50 &= 15 + 5 + 10 + 20 \\
\text{ATP}_5 &= \text{Total supply} - \text{Total demand} - \text{ATP}_1 + \text{ATP}_3 \\
60 &= 145 - 50 - 35
\end{aligned}$$

Thus the completed grid with ATP values added becomes as follows:

Part number: K7   Lead time: 3   Lot size: 60

	SOH	1	2	3	4	5	6	7	8
				DTF		PTF			
Independent forecast				20	20	20	20	20	20
Actual demand			15	5	10	20			
Proj. available balance	25	10	5	45	25	65	45	25	5
MPS (receipt)				60		60			
Available to promise			5		30		60		

The MPS system automatically recalculated the grid based upon any changes. For example, if the customer decides to place an order for the maximum recommended ATP quantity in week 1, the grid would be recalculated as follows:

Part number: K7   Lead time: 3   Lot size: 60

	SOH	1	2	3	4	5	6	7	8
				DTF		PTF			
Independent forecast				20	20	20	20	20	20
Actual demand			20	5	10	20			
Proj. available balance	25		5	0	40	20	60	40	20	0
MPS (receipt)				60		60			
Available to promise				30		60			

If the customer also decides to place an order for the maximum recommended ATP quantities in weeks 3 and 5, the grid would be recalculated as follows:

Part number: K7   Lead time: 3   Lot size: 60

	SOH	1	2	3	4	5	6	7	8
				DTF		PTF			
Independent forecast				20	20	20	20	20	20
Actual demand		20	5	40	20				
Proj. available balance	25	5	0	10	−10	30	10	−10	−30
MPS (receipt)				60		60			
Available to promise		0		0		60			

The negative figures in the PAB row would trigger action messages for the planner. These would be to reschedule the order for week 5 into week 4 and to raise a new order for week 7.

18.5 Two level MPS techniques

There are a wide variety of approaches to MPS and there is scope within this book to consider only general principles. In make to order (MTO), make to stock (MTS) and engineer to order (ETO) companies, the MPS is defined in terms of saleable products. This approach is relatively straightforward, though sometimes items are grouped into families to aid visualization of the MPS.

Many companies adopt a configure to order (CTO) approach. This has the advantage of allowing inventory to be held at the point of minimum variety. This combines the low stock and high variety of MTO/ETO with the short lead time of MTS. Master scheduling of CTO companies, however, is more complex. In this case, the MPS is usually defined at *two* levels. The first level relates to saleable product and is sometimes referred to as a final assembly schedule (FAS). This is driven by customer orders. The second level is based on forecast demand for lower level components or assemblies. Consider the diagrams in Figure 18.5.

If the products shown are machined from castings (shaded), then four types of raw materials will need to be held. If, however, a generic casting is used and then machined to order, only one type of raw material needs to be stocked.

The concept of a two level MPS is commonly applied in the automotive manufacturing industry (see Figure 18.6). In this case, the level 2 MPS is defined in terms of fictitious planning items. Figure 18.7 shows a so-called planning Bill of Material for a car.

Note that this bill is generic and does not represent a real vehicle. In order that components are ordered in realistic quantities, percentages are applied to the components to reflect usage. It is fundamental to this

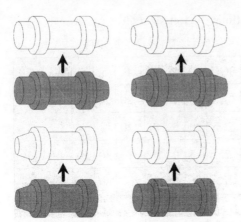

Figure 18.5 *Example illustrating two level MPS*

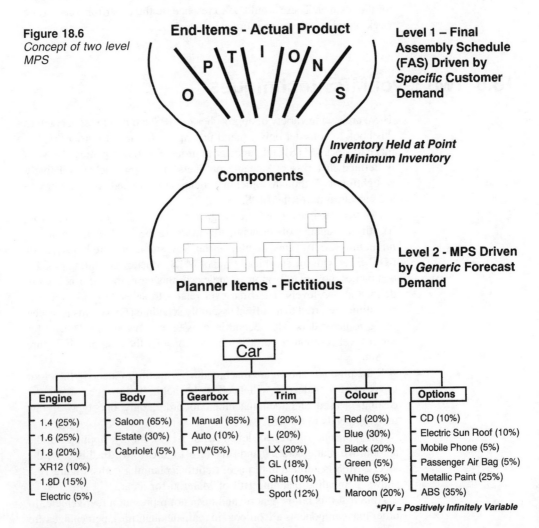

Figure 18.6
Concept of two level MPS

End-Items - Actual Product

OPTIONS

Level 1 – Final Assembly Schedule (FAS) Driven by *Specific* Customer Demand

Inventory Held at Point of Minimum Inventory

Components

Planner Items - Fictitious

Level 2 - MPS Driven by *Generic* Forecast Demand

Car					
Engine	**Body**	**Gearbox**	**Trim**	**Colour**	**Options**
1.4 (25%)	Saloon (65%)	Manual (85%)	B (20%)	Red (20%)	CD (10%)
1.6 (25%)	Estate (30%)	Auto (10%)	L (20%)	Blue (30%)	Electric Sun Roof (10%)
1.8 (20%)	Cabriolet (5%)	PIV*(5%)	LX (20%)	Black (20%)	Mobile Phone (5%)
XR12 (10%)			GL (18%)	Green (5%)	Passenger Air Bag (5%)
1.8D (15%)			Ghia (10%)	White (5%)	Metallic Paint (25%)
Electric (5%)			Sport (12%)	Maroon (20%)	ABS (35%)

*PIV = Positively Infinitely Variable

Figure 18.7 *Planning bill of material for a car*

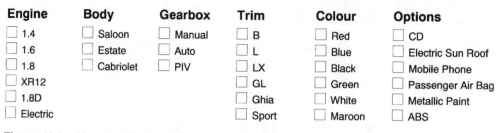

Engine	Body	Gearbox	Trim	Colour	Options
☐ 1.4	☐ Saloon	☐ Manual	☐ B	☐ Red	☐ CD
☐ 1.6	☐ Estate	☐ Auto	☐ L	☐ Blue	☐ Electric Sun Roof
☐ 1.8	☐ Cabriolet	☐ PIV	☐ LX	☐ Black	☐ Mobile Phone
☐ XR12			☐ GL	☐ Green	☐ Passenger Air Bag
☐ 1.8D			☐ Ghia	☐ White	☐ Metallic Paint
☐ Electric			☐ Sport	☐ Maroon	☐ ABS

Figure 18.8 *Modular bill of material for a car*

technique that these percentage splits on the planning bill of material vary little, even when there are variations in overall levels of sales. These are referred to as *planning bills of materials*. The level 2 MPS is driven by forecast. The result of applying planning bills is to generate a kit of items that will be consumed in the assembly of saleable product.

The level 1 MPS is driven by actual customer requirements and is sometimes called a final assembly schedule. This generates a conventional bill of material that is used to drive manufacturing. Note that this bill of material represents a vehicle (Figure 18.8) and because of its nature, is sometimes referred to as a *modular bill of materials*. It is inevitable that the forecast of planning bill items will not match actual requirements. Planning bills will, however, order items in matched sets.

18.6 Management aspects of MPS

It can be seen that MPS is the mechanism for management control of the organization over the short/medium term. Certain management structures are therefore essential for its successful operation. The traditional functional departments may well have different objectives:

- **Sales.** They will require high stocks and high variety to maximize opportunities for sales.
- **Production.** They will find low variety and large lot sizes convenient to manufacture.
- **Purchasing.** They will want to procure materials in large quantities to minimize costs.
- **Engineering.** They will want the best technical solution.
- **Finance.** They will require low stocks, high sales and low manufacturing costs.
- **Logistics.** They will require the minimum of disruption to simplify planning.

It is essential that all of the organization accept the MPS as the main business driver. In companies with successful master scheduling, it is usual to run a weekly MPS meeting involving all of the key company managers. It is essential senior management participates and supports this process.

To facilitate the MPS process, many companies appoint a master production scheduler. The responsibilities of the master scheduler are as follows:

- To maintain the MPS.
- To provide management with information.
- To ensure the MPS reflects the sales and operations (or aggregate) plan.
- To ensure the MPS is feasible.

Choosing the right person to be the master scheduler is crucial. Since the master scheduler is involved with the different functional departments, communication skills are essential. It is also important that the master scheduler has sufficient product/business knowledge and seniority to be credible. It is a common mistake to choose a very junior member of staff who is not taken seriously within the organization.

18.7 Summary

MPS is the tool for balancing supply and demand and as such is the mechanism for the control of the organization. Management uses MPS to define overall priorities which MRP can translate into a detailed manufacturing/purchasing plan. The MPS must have two attributes:

- It is feasible.
- It is sufficiently stable so as to prevent lower level change becoming unworkable.

For MPS to work, management commitment is essential. MPS cannot solve problems: it simply attempts to provide a feasible and coherent top-level plan. On occasion therefore it is likely that the MPS process will require difficult decisions to be made.

Finally, it should be noted that the ideas behind MPS are fundamental to all manufacturing management techniques. Certainly, MPS is a key element of JIT manufacturing. It is also worth pointing out that the MPS is simply a formalization of a process that must take place. In practice, if a particular plan is invalid, someone in the organization will make a decision (albeit in isolation). Without MPS, however, it is unlikely that this decision will reflect the true priorities of the whole organization.

19 Finite and infinite capacity planning

19.1 Overview

So far, planning of material requirements has been discussed. MRP systems generate planned works orders necessary to meet the needs of the company. So far, however, the issue of whether sufficient capacity is available to complete these orders has only been touched upon.

This chapter examines the mechanics of capacity planning. MRPII uses an infinite capacity planning approach, referred to as capacity requirements planning (CRP). Finite capacity scheduling (FCS) is also available and this will also be described. Finally, rough-cut capacity planning (RCCP), which is a simplified approach used to support MPS, will be outlined.

19.2 Limitations of MRPII

MRP is the scheduling engine within MRPII. It is based on two assumptions:

- The vast majority of MRP systems use fixed lot sizes. While algorithms exist for part-period balancing (i.e. use of variable lot size), these are only rarely applied.
- More significantly, all MRP systems use fixed lead times. In effect, this assumes there is no interaction between different works orders.

The limitations of MRP are illustrated in Figure 19.1. Consider a manufacturing system with two workcentres. Two different items, A and B, are manufactured according to the routing shown. The total time to make a batch of A is 11.33 hours. The total time to manufacture a batch of B is 12.17 hours. Applying a simple back scheduling logic (with no allowance for queue), B would be manufactured first. Because item A requires a relatively short time to complete operation 10, intuition would suggest that it should be manufactured first.

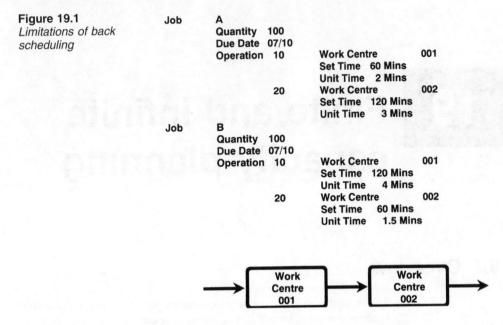

Figure 19.1
Limitations of back scheduling

Job	**A**
	Quantity 100
	Due Date 07/10
	Operation 10

Job A
Quantity 100
Due Date 07/10
Operation 10   Work Centre   001
              Set Time   60 Mins
              Unit Time   2 Mins
           20   Work Centre   002
              Set Time   120 Mins
              Unit Time   3 Mins

Job B
Quantity 100
Due Date 07/10
Operation 10   Work Centre   001
              Set Time   120 Mins
              Unit Time   4 Mins
           20   Work Centre   002
              Set Time   60 Mins
              Unit Time   1.5 Mins

While this is a contrived example, it illustrates the fact that simple MRP back scheduling does not consider the interaction between different batches of work. As has been shown earlier, however, in the vast majority of cases, the largest element of manufacturing lead time is queue. As the value-adding element of lead time increases, then the assumptions underpinning MRP become less and less valid. For MPS to work effectively, therefore, it is essential that WIP levels remain relatively constant.

19.3 Infinite vs finite capacity scheduling

At a fundamental level, there are two forms of scheduling: infinite and finite. Infinite capacity scheduling (ICS) underpins MRP systems. In ICS, the due date of jobs is calculated by subtracting a fixed lead time from the required date. In principle, this can lead to start dates being calculated that are earlier than the current date. If a job has a number of operations, the starting date of each is determined by apportioning the total lead time.

There are several techniques for accomplishing this, but typically, the lead time is divided equally between the operations. Because start dates are calculated by subtracting a lead time from a due date, ICS is a *backward scheduling* technique.

The term 'infinite' in the context of scheduling refers to the fact that start dates are calculated without regard to capacity (i.e. an assumption of infinite capacity is made). In finite capacity scheduling (FCS), capacity is considered *during* the scheduling process (i.e. finite capacity is assumed). Typically, FCS works *forward* from the current date to

Figure 19.2
Backward scheduling

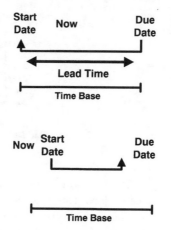

Figure 19.3
Forward scheduling

calculate start and due dates. Thus FCS is usually referred to as a *forward scheduling* technique. Note that in FCS, the finish date of the job may be before the due date (see Figure 19.3) or after (i.e. if the job is late).

19.3.1 Infinite capacity scheduling

The process of ICS is best explained by an example. Consider a small machine shop with six workcentres or resources (R1–R6) and six jobs. The routing for the six jobs defines the path through the machine shop. In this case, all of the jobs require four operations. The mode of production is discrete batch manufacture. The length of time required for any operation for job *j* and operation *i* is given by the expression:

$$T_{Opij} = L_{sj} T_{pij} + T_{sij} \tag{19.1}$$

Where T_{Op} = length of operation, T_p = process time and T_s = set-up time.

Using equation (19.1), the length of each operation can be calculated and this is represented in Figure 19.4. Note that the length of each 'bar' in the Figure is proportional to T_{Op}.

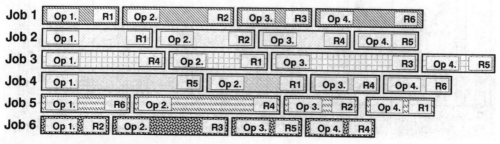

Figure 19.4 *Routing example*

Figure 19.5   *Back scheduling example*

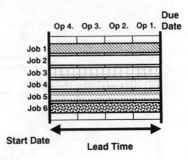

The process of ICS starts with determining the start date for each of the six jobs. This is accomplished by back scheduling from the due date (in this case, this is assumed to be the same for all of the jobs). If a four-day lead time is assumed, all of the jobs will start at the same time.

The next stage is to determine the start date of each operation. The commonest method of doing this is to apportion the lead time equally between the operations. Thus, for the first job, the first operation will start at the beginning of day 5 and be completed at the end of the same day. This back scheduling process will continue until all of the operations for all of the jobs have been calculated. This is shown in Figure 19.5.

At this point, capacity has not been considered. It is perfectly possible that one or more of the resources will not have the capacity to complete the schedule. Taking resource 1, the total load profile for the entire period can be determined by adding the individual T_{Op} values for the individual days. This is illustrated in Figure 19.6.

Note that the available capacity for this resource is just sufficient to satisfy the imposed load. If the process is repeated for all of the resources, however (see Figure 19.7), it can be seen that there is an overload on resource 2 on day 5 and resource 3 on day 7.

There are three key points associated with ICS:

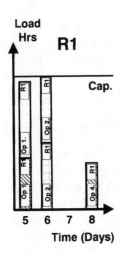

Figure 19.6   *Load profile for R1*

- With ICS, jobs are never late *by definition*.
- Problems with the schedule are shown by *overloads*.

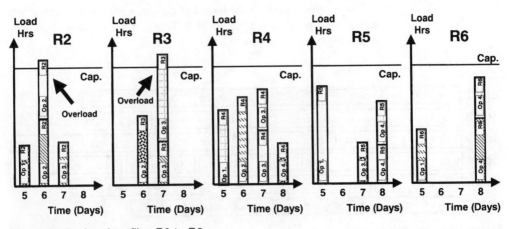

Figure 19.7   *Load profiles R2 to R6*

- ICS provides management with information of when overloads will arise. It does not, however, provide any advice on how the overload can be addressed. Management needs to decide therefore if capacity will be added (e.g. with overtime) or a job will need to be rescheduled.

19.3.2 Workcentres and routings

It should be noted that to support scheduling data relating to workcentres and routings must be available. These are described below.

Workcentres. A workcentre is defined as an entity where a works order is processed. In most cases, the terms 'workcentre' and 'resource' can be used interchangeably. Typically, a workcentre will have a defined level of capacity. Normally, workcentres are defined at a more detailed level than a resource. Typically, a workcentre will be an individual machine or a group of machines. Workcentres may also be defined in terms of labour. The following information is associated with a workcentre:

- Workcentre code
- Number of operators/machines in group
- Efficiency factor
- Workcentre description
- Available hours per shift
- Cost rate

Routing. A routing defines which workcentre an item visits during manufacturing. Thus a routing consists of a number of *operations*. Each operation is associated with the following information:

- Sequential operation number
- Set time
- Move time
- Operation description in text form
- Workcentre code
- Process time
- Queue time

19.3.3 Finite capacity scheduling

There are several FCS software packages available on the market. It is possible, however, to accomplish FCS using manual methods. Gantt charts, developed early in the 1900s, are effectively a graphical approach to FCS.

There are several approaches to FCS. All of these, however, have a number of features in common. FCS is best explained by a simple example. Consider the case described in 19.3.1. The first stage in scheduling the six jobs to finite capacity is to determine the priority

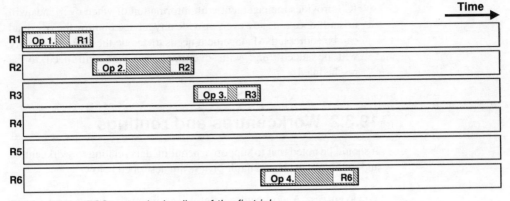

Figure 19.8 *FCS example: loading of the first job*

sequence (i.e. to determine which job is most important, which is second and so on). Assuming the priority sequence is job 1, job 2, job 3, job 4, job 5 and then job 6, this is the order in which the jobs will be loaded (methods of establishing priorities will be discussed in a later section). Finally, the schedule can be optimized.

The first stage is to create a grid for the six resources. The highest priority job (job 1 in this case) can be loaded as shown in Figure 19.8.

Notice that the only constraint on the completion of operations is the time required for earlier operations. When the second job is loaded, however (Figure 19.9), there is interference. The first operation of the second job cannot be scheduled immediately onto R1 because it is occupied with the first job.

The process can be continued until all of the operations for all jobs have been loaded. The final result is illustrated in Figure 19.10. There are three key points associated with FCS:

- Resources are *never* overloaded by definition.
- If there is insufficient capacity to produce the complete schedule, this is shown by one or more jobs being scheduled late (i.e. the scheduled due date is later than the required date).

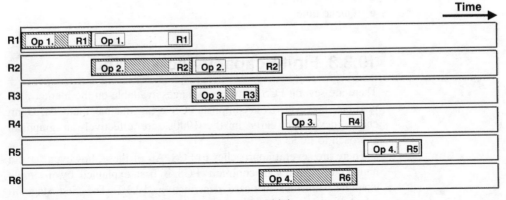

Figure 19.9 *FCS example: loading of the second job*

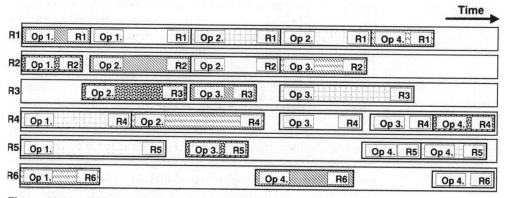

Figure 19.10 *FCS example: all jobs loaded*

- If any jobs are completed late, these are selected purely by the priority rules employed. Any subjective factors, such as the importance of any customer at any moment of time, are ignored.

For realistic numbers of resources and jobs, the use of a computer is obligatory.

19.4 Schedule optimization

An examination of Figure 19.10 shows that some start and finish times of some operations for some jobs are not simultaneous. This indicates a queue of work. It can also be seen that the grid has gaps where no job operation is scheduled. This indicates that a resource is idle. The question naturally arises therefore, could the schedule be adjusted to take advantage of the gaps and queues to produce a 'better' schedule? That is to say, could the schedule be optimized? As before, it is necessary to define an objective function to allow optimization to be performed. Possible objective functions include:

- **Minimum makespan.** This minimizes the total time to produce all of the jobs.
- **Minimum tardiness.** This minimizes the aggregate 'lateness' of the jobs.
- **Maximum profit.** This will have the effect of finishing highly profitable jobs in preference to others.

One way in which optimization could be accomplished would be to try *all* possible schedules and select the one that gave the best results in terms of the objective function. It can be shown, however, that the number of possible schedules for m resources and n jobs is given by m^n. For typical values of m and n, even powerful computers cannot evaluate all of the permutations in realistic time periods. Techniques, such as branch and bound methods, can be used to give *optimal* solutions. These are, however, beyond the scope of this book. Nonetheless, a relatively simple approach does exist for problems with two resources and *n* jobs. This is known as Johnson's algorithm.

19.4.1 Johnson's algorithm

Johnson's algorithm dates back to the mid-1950s. It is illustrated as shown in the flow chart in Figure 19.11. The method is best explained by an example. Consider Table 19.1. This shows ten jobs (A–J) that need to be scheduled through two machines. All of the jobs first need to visit machine 1 and then machine 2. The times shown in the table relate to the time that it takes to process the work at each machine.

Figure 19.11
Johnson's algorithm

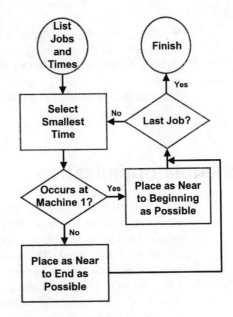

Table 19.1   Example for Johnson's algorithm

	Machining time	
Job	Machine 1	Machine 2
A	6	7
B	5	4
C	11	5
D	4	8
E	9	12
F	10	13
G	13	9
H	7	10
I	8	9
J	8	6

Table 19.2 Completed example for Johnson's algorithm

Job	Machining time Machine 1	Machine 2	Seq.
A	6	7	D
B	5	4	A
C	11	5	H
D	4	8	I
E	9	12	E
F	10	13	F
G	13	9	G
H	7	10	J
I	8	9	C
J	8	6	B

The first stage is to define a one-dimensional array of extent ten (one element for each job). The first job that will be selected is B. Because the smallest value occurs on the second machine, this will be allocated to the tenth position. The next job to be selected will be D. In this case, the minimum occurs at the first machine. Thus it will be allocated to the first position. This process is repeated until all of the jobs have been allocated. The final schedule calculated is as shown in Table 19.2.

19.5 Sequencing

19.5.1 Overview

There are no optimum solutions to the three-machine problem. It is important nonetheless to provide some discipline to job scheduling. This is typically achieved by the use of a *sequencing rule*. These rules are used by the manager of a resource to answer the simple question, which job in the queue should be processed next?

19.5.2 Sequencing rules

Sequencing rules can be divided into three broad categories:

Static. Here the priority of the job remains constant over time. Examples include earliest due date (EDD), shortest processing time (SPT) and longest processing time (LPT).

Dynamic. In this case, the priority is continually changing as time passes. Because of this, it is necessary to undertake a calculation for all of the jobs in the queue when a priority decision is made. The most common rules are first-in-first-out (FIFO), critical ratio (C_r), least slack

(L_{slack}) and least slack per remaining operation (L_{SPRO}). These will be discussed in more detail later.

Informal. Priorities are determined in this case by informal rules applied by operators, first-line supervision or progress chasers. These rules include easiest next set-up, least loaded downstream machine. Often, the rules applied are difficult to establish, even with detailed investigation.

Choice of an appropriate sequencing rule is essential. In most factories, the most significant element of manufacturing lead time is queue. If jobs are not prioritized logically, then some jobs will be completed late and others early. In order to maintain good flow and achieve reasonable due date performance, a logical rule should be chosen. The most commonly applied rules are discussed below:

FIFO. This is the simplest rule to apply. It can be implemented physically by placing a conveyor in front of the resource. Because it does not take into account the required date of the order, however, it can lead to some jobs being selected in preference to those that common sense would suggest are more urgent. If excess capacity is available, however, the simplicity of this rule makes it attractive.

EDD. Again, a relatively simple rule – the job selected is simply the one with the earliest due date. The only flaw with this rule is that it does not take into account the work content of the jobs in the queue.

SPT. Under most circumstances, the practice of selecting the shortest job has a clear flaw; the longest jobs never get completed. Under conditions of extreme overload, however, there is some logic in completing the maximum number of jobs.

C_r. This is a more sophisticated rule: it is the ratio of the time required to the time available (T_a):

$$C_r = \frac{\sum_{i=m}^{i=n} (L_s T_{pi} + T_{si})}{T_a} \tag{19.2}$$

where n = total number of operations and m = current operation.

Note that if the job is overdue, the critical ratio becomes negative. This can cause problems if a computer algorithm is used to sequence jobs.

L_{slack}. Slack is defined as the difference between the time available and the time required:

$$\text{Slack} = T_a - \sum_{i=m}^{i=n} (L_s T_{pi} + T_{si}) \tag{19.3}$$

Again, note that if the job is overdue, slack becomes negative.

L_{SPRO}. Slack per remaining operation (SPRO) overcomes a flaw of the L_{slack} rule. As previously stated, in most cases, the dominant element in lead time is queue. Thus, if two jobs have the same slack, the one with the greater number of operations left should be selected. Thus SPRO can be calculated as illustrated in equation (19.4).

$$\text{SPRO} = \frac{T_a - \sum\limits_{i=m}^{i=n} (L_s T_{pi} + T_{si})}{n - m} \tag{19.4}$$

Overall, the L_{SPRO} rule gives the best results under normal conditions of operation.

Easiest next set-up. If applied indiscriminately, this rule will create difficulties. Jobs with difficult set-ups will be neglected and in the long term, due date performance will suffer. If, however, a resource is forced to manufacture products in a particular sequence (L_{SPRO}, for example), this might also cause problems. In practice, *set-up time dependency* is common: that is to say a particular sequence of manufacture can reduce set-up time. The most obvious example of this effect is in painting processes. Clearly, it is more convenient to paint different colours in a sequence starting with the lightest shade, working through to the darkest. To overcome this apparent contradiction, one technique that has been applied is to allow a resource a 'window' of a fixed number of jobs. This is best explained by an example. Consider Figure 19.12. There are ten jobs queuing (shown in priority order). The resource is given a window of five jobs. This means that the resource is authorized to select jobs A, B, C, D or E. Before job F could be chosen, however, it would be necessary first to complete job A. This approach provides a compromise between the conflicting demands of the need to manufacture in priority sequence and the constraints imposed by set-up time dependency.

It should be noted that overall scheduling performance is insensitive to the application of sequencing rules; provided the rules chosen are reasonable, the overall business outcome is more dependent on appropriate levels of capacity being available. This is why the technique described in Figure 19.12 is effective. It should also be noted that scheduling effectiveness is insensitive to the calculations for priority values. It is not necessary therefore to know the values of run-time and set-up to high levels of accuracy.

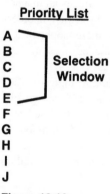

Priority List

A
B
C
D
E
F
G
H
I
J

Selection Window

Figure 19.12
Priority vs set-ups

19.6 Short interval scheduling

A relatively recent development is the introduction of short interval scheduling (SIS) systems. These are FCS systems designed to operate alongside MRPII. These systems are illustrated in Figure 19.13.

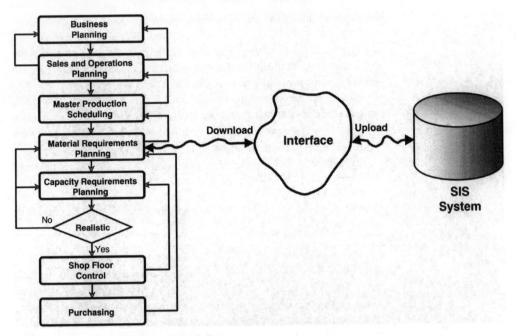

Figure 19.13 *Short interval scheduling*

An MRPII system is used to create works orders in the usual manner. These are downloaded via an interface into the SIS system. The orders are then forward planned and in some of the more sophisticated versions, also optimized. The orders are then uploaded back into the MRPII system.

This hybrid approach uses the hierarchical planning logic of MRP while giving finer control over shop floor operations. SIS can also be used for sophisticated 'what-if' simulations. This can be used at a detailed level to determine the effect of, say, taking a machine out of service for preventive maintenance. Alternatively, it can be used to supplement or even replace RCCP in the master production scheduling process.

19.7 Problems of finite capacity scheduling

FCS appears to be a more scientific approach to manufacturing management than MRP. There are a number of shortcomings, however:

- FCS is complex, both conceptually and in the operation of the software. FCS requires a considerable amount of data to be collected. The management effort to implement these systems can be considerable.
- FCS software in general, and optimized production technology in particular, can be very expensive.

- Many processes in manufacturing industry are highly stochastic. This can undermine the precision of the FCS process.
- Set-up time dependency can also cause the base data for FCS to be unreliable. While tables can be provided for set-up time dependency, the amount of data required can be considerable.
- FCS systems cannot create capacity. As was discussed earlier, FCS systems do not necessarily maintain the integrity of sales order due dates. Without management appreciation of the inherent limitations of FCS, it is easy to overlook the fundamental issues of master scheduling.

Overall, FCS is vulnerable to many of the same problems that can be encountered with any systematic approach to manufacturing management. FCS is not a substitute for effective management.

19.8 Feedback and work to lists

To implement any scheduling system, there are two requirements. First, to gather information from the factory such that the status of orders on the shop floor is known. Second, to provide detailed information to each workcentre on the relative priority of the jobs awaiting production.

19.8.1 Shop floor feedback

After each operation as defined in the routing, it is necessary to report back to the system. The information that typically needs to be reported is as follows:

- Number of products successfully completed. This may change the quantity to be received on a particular order so that the MRP logic may prompt action messages. This also permits the progress of particular work orders on the shop floor to be monitored.
- The time required to complete the operation. This is reported in some systems in order to compare the planned time as defined by the routing with the actual time taken.

There are a number of techniques for shop floor feedback:

- **Remote entry.** This is the traditional approach to shop floor feedback. Shop floor personnel would write the necessary information on paper to be keyed into a computer system by a specialized data entry department. This approach can create transcription errors, however, and also means that responsibility for data accuracy cannot be clearly defined.
- **Shop floor terminals.** This approach is now popular as the cost of computer terminals has fallen.

- **Data collection.** Automated approaches to data collection can be used. Methods include:
  - Bar code readers
  - Hand held terminals
  - Smart tags
  - Direct connection of plant

19.8.2 Work to lists

The output from any detailed scheduling approach is the *work to list*. The objective of this document is to define the sequence in which a particular workcentre should process orders. A typical work to list is shown in Figure 19.14.

Work Centre	0290 Centre Lathes						Issue Date 17/04/05

Work Order Number	Part Number	Quantity	Operation Due Due Date	Order Due Date	Next Work Centre	Previous Work Centre
3000167	GOS1365	1000	18/04/05	06/05/05	0360	0370
3000369	GDS1987	1000	18/04/05	07/05/05	0360	0370
3000258	PFF1888	1000	18/04/05	12/05/05	0380	0410
3000198	TDS3687	1000	18/04/05	16/05/05	0360	0370
3000153	TDS9637	1000	19/04/05	17/05/05	0360	0420

Figure 19.14 *Example of a work to list*

19.9 Rough-cut capacity planning

19.9.1 Overall concept

Rough-cut capacity planning (RCCP) is used after MPS to verify that the plan is valid. It is far simpler than the scheduling approaches discussed so far. RCCP is used in conjunction with MPS before MRP is run. It therefore does all levels in the bill of materials. It is necessary therefore for simplifying assumptions to be made.

RCCP requires one or more *generic resources* to be defined. A generic resource is a capacity constraint and is typically defined more broadly than a workcentre. It can take a number of forms:

- Production line.
- Critical workcentres.
- Labour.
- Key supplier capacity.
- Testing.

RCCP also needs a *product load profile* for each master scheduled item. This defines how much load per unit produced is imposed on the RCCP resources. This is a time-phased plan and a particular resource may be required several times during the manufacture of a product. The use of product load profiles in complex products represents an approximation of the actual load.

The data is aggregated for all master scheduled items to produce a time-phased load profile for each of the resources defined. Where overloads are detected, management need to take appropriate action. Preferably, this should take the form of additional capacity, as changes to the MPS will cause instability at lower levels in the planning process.

The use of RCCP allows the impact of a particular product mix with the MPS to be analysed. It allows 'what-if' analysis so that before the MPS is submitted to MRP, management can be reasonably sure that the plan is broadly compatible with capacity.

RCCP does have limitations nonetheless:

- It considers only critical resources. This is not a problem unless product mix is highly variable.
- It ignores WIP/component inventory of lower-level components.
- It does not take into account the completion of work orders.
- The effect of these simplifications is that RCCP is a pessimistic view of the capacity/load balance.

In practice, RCCP is often a critical element in successful master scheduling. If skilfully applied, the limitations outlined do not invalidate the technique.

Some software suppliers now offer more sophisticated RCCP systems based on simulation techniques. These provide a more accurate picture of load and help to visualize the effect of product mix.

19.9.2 CRP vs RCCP

CRP and RCCP have the same objective: to provide information on the load imposed on the manufacturing system. Both are *infinite capacity loading* approaches. That is to say, both maintain works order start and finish dates. It has been shown that these approaches contrast sharply with FCS where capacity is not exceeded, but due dates are flexible. Table 19.3 compares these two techniques.

A superficial review of these two techniques would suggest that CRP is the superior technique. CRP has three major disadvantages, however:

Data. CRP requires a large amount of data. With the decline of bonus systems, many companies no longer have work study departments to provide accurate routing information.

Need for feedback. To provide an accurate answer, CRP requires detailed shop floor feedback. Many companies are attempting to reduce the amount of transactions that need to be made on the shop floor.

Timing of information. CRP is run after MRP. Thus, if any significant overloads are detected that cannot be addressed by overtime, then it will

Table 19.3 Comparison of RCCP and CRP

	RCCP	CRP
Definition	Estimated load on critical resources based on MPS	Detailed evaluation of load based at workcentre level
Method	Use of MPS and product load profiles	Calculation based on all works orders
Frequency	As required	Following each MRP run – typically once a week
Objective	1 Pre-MRP evaluation of MPS 2 Operational planning	1 Post-MRP analysis 2 Determining bottlenecks
Precision	Aggregate	Detailed
Data	MPS and product load profiles	Works orders, workcentres, routings and works order status
Speed	Fast	Typically longer than MRP
Use	Virtually all users of formal manufacturing management	A minority of users

be necessary to change the MPS. As has been discussed earlier, this has serious implications for MRP planner action lists.

With the increasing sophistication of RCCP systems, CRP is now used by only a minority of organizations. Nonetheless, the concepts underpinning CRP are important to understand. Moreover, the techniques underpinning CRP are a component of optimised production technology (OPT) as will be discussed in 19.10.

19.10 Optimized production technology

As stated previously, many practitioners have expressed doubts over the use of MRPII. The success of Japanese manufacturers suggested a more pragmatic approach to manufacturing management. In the 1970s, an Israeli physicist, Dr E.M. Goldratt, developed an apparently new approach to manufacturing management called optimised production technology (OPT). OPT can be considered to have two parts: first, a philosophical element, which outlines a new approach to manufacturing management. Second, a finite scheduling algorithm that is more complex than MRP and is claimed to be more effective.

19.11 OPT principles

OPT is derived from a simple principle: that it is the goal of a manufacturing organization to make money. In financial terms, this can be represented by three measures:

- Net profit.
- Return on capital employed (ROCE).
- Cash flow.

These measures can only be determined at the highest level in the business and are difficult to relate to the operational level. OPT proposes three operational measures that can be directly related to the financial measures. These are defined very precisely:

Throughput. The rate of generating cash through sales. An increase in throughput will lead to an improvement in net profit, ROCE and cash flow.

Inventory. This is the value of everything that has yet to be converted into throughput. It should be noted that OPT does not consider the concept of added value, unlike traditional management accounting systems. A decrease in inventory will improve ROCE and cash flow.

Operating expense. This is the total cost of converting inventory into throughput. Note that OPT does distinguish between direct and overhead costs. A reduction in operating expense will improve net profit, ROCE and cash flow.

OPT takes these relatively simple principles and combines these with the observation that the vast majority of manufacturing systems are not perfectly balanced. In practice, output is constrained by a small number of critical resources, generally referred to as *bottlenecks* (or *capacity constraining resources (CCR) or critical resources*). A bottleneck is defined as any resource where the available capacity is less than the demand and thus constrains the system as a whole. OPT suggests that bottlenecks are fundamental in the planning and control of the manufacturing system. This is embodied in the so-called ten rules of OPT which contrast with traditional approaches:

1 **Balance flow, not capacity.** Traditionally, companies have attempted to maintain even utilization of resources, even if this has disrupted workflow.
2 **The level of utilization of a non-bottleneck is determined not by its own potential, but by some other constraint in the system.** Traditionally, each resource is planned and measured independently.
3 **Utilization and activation of a resource are not necessarily the same thing.** This implies that low efficiency of a resource is not necessarily a bad thing, which is fundamentally opposed to traditional management techniques where payment of direct labour is often based on output.
4 **An hour lost at a bottleneck is an hour lost for the total system.** This implies that management attention should be focused on these resources.
5 **An hour saved at a non-bottleneck is a mirage.** This is a consequence of rule 2.

6 **Bottlenecks govern both throughput and inventory.** Traditionally, bottlenecks were treated as short-term limitations and not fundamental to the system.

7 **The transfer batch may not, and many times should not, be equal to the process batch.** This rule promoted the idea of overlapping batches in order to reduce lead time.

8 **The process batch should be variable, not fixed.** This contrasts to the EOQ approach.

9 **Capacity and priority should be considered simultaneously, not sequentially.** In effect, this promotes the use of finite capacity scheduling techniques.

10 **The sum of local optima is not equal to the optimum of the whole system.** This supports the Japanese approach of considering the whole system.

19.12 OPT scheduling logic

Complementary to (but separate from) the OPT principles is the OPT software. This was originally developed in the early 1980s by a company called Creative Output.

Finite capacity scheduling systems have been available for several years. The problem with these systems is the complexity of the scheduling problem. In a factory with m machines and n jobs, the number of possible schedules is m^n. For realistic numbers, this is an intractable problem, even with the fastest computers.

OPT overcomes this problem by the observation that only bottlenecks need to be to planned carefully; non-bottlenecks are less important. The basic principles of the OPT planning logic are shown in Figure 19.15.

BUILDNET. Routing and BOM information is merged to give a combined network representing the factory and all its jobs.

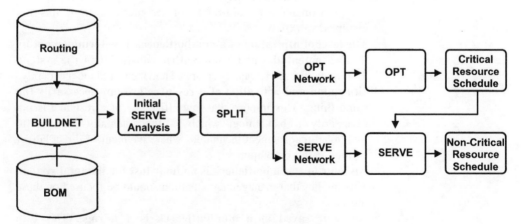

Figure 19.15 *Representation of the OPT scheduling process*

SPLIT. The system divides the network into critical and non-critical elements.

OPT. The critical network is scheduled to finite capacity. The algorithm behind this process is secret and proprietary. The algorithm determines a schedule for the critical resources and process and transfer lot sizes. OPT attempts to allow for set-up time dependency by providing for tables to be created that define the time required to change between different items.

SERVE. The non-critical network is scheduled to infinite capacity. Critical to these schedules is the requirements of the bottleneck resources.

Because of the relatively small number of implementations of OPT compared to MRPII, it is difficult to objectively compare the operational effectiveness of the two approaches. While some companies have reported success with OPT, some users have not obtained the anticipated benefits. This is an area that requires further research.

19.13 Summary

As the cost of computing has reduced, FCS is becoming more popular. It has evolved from early programs in the 1950s to a wide range of applications today. Some companies offer FCS in the form of enhanced RCCP systems to be used in parallel with MRPII. Others offer FCS in the form of short interval scheduling (SIS) systems as an aid to detailed shop floor planning. The OPT philosophy has proved extremely influential, but some of the implementations of the software have been perceived to be unsuccessful. While OPT is not attracting the same level of interest as when it was first proposed, the number of implementations is growing. The whole area of FCS is continuing to develop.

19.14 Case study

19.14.1 AlterCo – part 1

AlterCo Ltd is a manufacturer of alternators and DC motors based in East Lancashire, UK. It has 315 employees and a turnover of £15 M per year. The company produces motors for a wide range of customers and applications. The basic material flow is shown in Figure 19.16.

The product is made from three major components: a shaft, a winding and two end covers (see Figure 19.17). Weekly production of these (at the time of the case study) was 7800, 6250 and

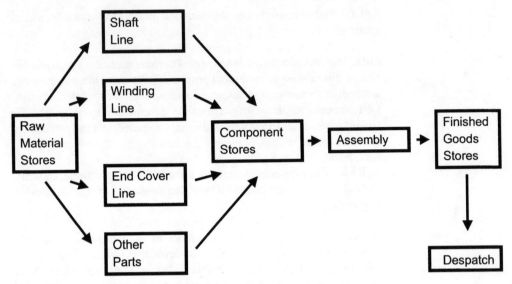

Figure 19.16 *Manufacturing flow for AlterCo*

Figure 19.17 *Bill of material for AlterCo*

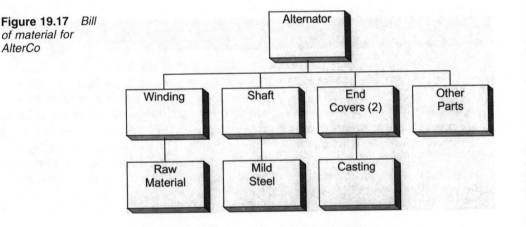

14 000, respectively (note, there are two end covers for each final product). The other components are either purchased or have a low variety of types. It was generally accepted that the winding process was the most difficult in the factory. Indeed, in production meetings, the point was often raised that winding consistently produced the lowest output figures of any dept.

The company had lost significant market share in the 1970s leading to many redundancies. During the 1980s, however, world markets had improved and the demand for the product had largely recovered. The company found itself in the unfortunate position of not being able to meet demand. The company had 21 000 units overdue and this figure was rising. Excessive overtime was being worked in the component manufacturing areas, though the assembly areas often ran short of work. Despite the fact that stocks of components were growing, a great amount of management effort was put into finding enough components to support assembly. The worst case was shafts: there were 35 000 shafts in components stores, yet there were continual shortages.

The company had a sophisticated MRPII system, but this had lost credibility. Data accuracy was poor, but manufacturing personnel were particularly critical of the reports produced by the system. These were very long and consisted mostly of lists of overdue parts. Some software modules (e.g. master production scheduling and capacity requirements planning) were not used.

The company's position was becoming increasingly serious, particularly in terms of cash flow. The initial response to this was to increase batch sizes in the component manufacturing areas in order to minimize the effect of set-ups. Each manufacturing foreman also sequenced the work through his particular area to make machine set-up easier. This was particularly common in the winding section where the nature of the process made sequencing particularly advantageous from a set-up point of view. These actions did not alleviate the problems, however. The manufacturing foreman complained that they were given insufficient lead time and said that if lead times were increased, this might allow them to reduce the number of overdues. The foreman also stated that they had insufficient capacity despite overtime working.

At this point, a project team was set up consisting of two personnel from the company and an employee from a company in the same group. Their objective was to make short-term improvement to the serious situation at AlterCo.

What was the team's assessment of the problems and what action did they take?

19.14.2 AlterCo – part 2

The team recognized almost immediately that winding was a bottleneck. What was surprising was the fact that the other sections (particularly shafts) themselves had a large number of overdues. The reason for this was that the other sections were making the wrong types. For example, in order to improve efficiency within the shaft section, jobs were produced in a slightly different sequence to that recommended by the MRP system. This led to an unmatched shaft and winding entering stock. Since these could not be matched, assembly could not produce the final unit. This led to overdue units and hence overdue shafts. Thus, there was even more incentive to alter the manufacturing sequence to gain greater efficiency in the shaft section.

To add to this problem, once the manufacturing sections began to fall behind, they inflated batch quantities with the idea of improving efficiency. The effect of this was to put too much of one type into stock and not enough of another. Again, this led to stocks increasing and overdue lists growing. Overtime simply made problems worse – more of the wrong parts were made. What was not recognized by the company was that only one of the manufacturing sections was a capacity constraint. With the benefit of hindsight, this appears obvious. The production figures showed that shafts consistently outproduced winding by 1550/wk (i.e. stock of shafts was increasing by 1550/wk).

Since the manufacturing sections had large numbers of overdues, they asked for longer lead times. This led to material being released too early and work in process (WIP) increasing. Now lead time is largely determined by queue lengths. As the lead time was increased, so queue lengths went up. As before, this process is self re-enforcing.

Finally, because each of the manufacturing sections was making the wrong parts, obtaining full kits for assembly became increasingly difficult. This meant even the stock of windings increased.

The short-term solution was to make the other manufacturing sections produce matching parts to the winding section. As jobs were released onto the winding line, a note of the part number of the appropriate motor was taken. The computer-based bill of material (BOM) on the MRP system was then used to break the final product down into its constituent components. The other components were then netted off against stock and this constituted the schedule. Since winding was the bottleneck, they were given some freedom in setting their schedule; that is to say, the winding foreman was allowed to sequence work through the section. As far as possible, however, the winding foreman tried to work in the sequence suggested by the MRP schedule (i.e. most overdue jobs first). Effectively, the MRP schedule drove the winding section and it drove the rest of the factory as shown in Figure 19.18.

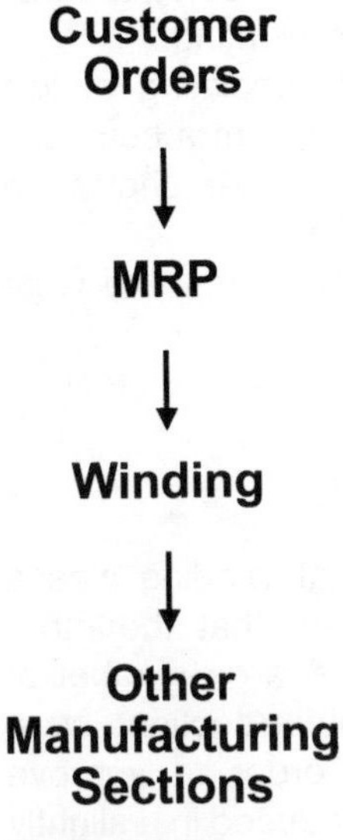

The immediate result was that the other manufacturing sections were 'starved' of work (since there were massive stocks to be consumed). This led to overtime being stopped on all sections except winding. Even so, the other sections ran short of work and it was suggested by the team that the excess of capacity could be used to set up more times and therefore produce smaller batches. This caused protest among the foremen since their efficiencies would drop (did this matter?). One foreman stated that he felt it was *unfair* that winding should be given more freedom. Protest was so strong that the project was almost discontinued. Fortunately, benefits started to become apparent:

Customer Orders

↓

MRP

↓

Winding

↓

Other Manufacturing Sections

Figure 19.18
AlterCo's solution

- Output of finished products increased because it was easier to match up complete kits of parts. The efficiency of assembly also went up. It was also found that the non-bottleneck areas had sufficient capacity to make matching parts to existing stocks of windings. Because of this, the *initial* output of finished products was actually greater than the output of the winding section.

- The most overdue orders were being produced first. When management was 'scrambling around' to find complete kits for assembly, they were often *not* the most urgent jobs.
- Costs went down due to reduced overtime working.
- Since the large inventory of components was being 'worked off', purchasing did not need to buy so much raw material. In the short term, this improved the company's cash flow.
- Because WIP was reduced, so were lead times. This made manufacturing more responsive.

In numerical terms, the benefits can be summarized as shown in Table 19.4.

Table 19.4 AlterCo's performance

Measure of performance	January	April
Shaft stock	35 000	25 000
Shaft WIP	15 000	6400
Shaft lead time	2.5 wks	1 wk
End cover stocks	40 000	32 000
Shaft output	7600	5010
Overall output	6000	8700
Overall overdues	21 000	13 000

In the short term therefore, the improvement project was considered to be a great success. The project team advised caution, however. Why was this and what were their recommendations for further improvements?

19.14.3 AlterCo – part 3

The project team was very concerned that the short-term system only worked due to the large inventories being worked off. While stocks were being consumed, it was possible to run non-bottleneck workcentres very inefficiently. It was realized, however, that once the large stocks had been consumed, it would be necessary to improve the efficiency of non-bottlenecks (though not to 100%). Moreover, since the winding section was driving the rest of the factory, the other manufacturing sections got very little advance warning of production requirements. This made planning very difficult.

The team's recommendation, therefore, was that production planning was carried out at end product level taking into account the capacity limitations of winding. A capacity planning system could then be used to check if the plan was valid (looking at all parts of the factory), but paying particular attention to winding.

19.14.4 Analysis

In summary, the actions taken by the project team seem obvious with the benefit of hindsight. In an industrial environment, however, day-to-day pressures can often obscure the seemingly obvious. It is essential therefore to look for the root cause of problems, rather than attacking the symptoms. It is all too easy to increase overtime, batch sizes, lead times and stocks when the real problems lie in ineffective production coordination or poor system design.

Note also that the problems of this organization could not be solved by an examination of the individual processes. Indeed, this approach made the situation worse. In this case, it is necessary to consider manufacturing as an integrated system.

20 Just-in-time

20.1 Overview

By the early 1980s, MRPII had been adopted by many manufacturing companies. The early success of organizations implementing MRPII, however, had not be emulated more widely. One of the major figures in the development of MRPII was Oliver Wight. He developed a simple classification system to measure the effectiveness of MRPII. Surveys suggested that only a very small proportion (<10%) achieved the highest grade on the classification system, class 'A'. At the same time, the spectacular success of Japanese manufacturing companies was becoming obvious in the west.

These factors led to a re-evaluation of MRPII and its underpinning assumptions. This chapter will review some of the manufacturing management techniques that originated in Japan and Toyota in particular. These ideas are often referred to as just-in-time (JIT) manufacturing.

20.2 Japanese manufacturing practice

It is dangerous to summarize JIT as a concept because of its breadth, but perhaps the most applicable definition is as follows:

> *An approach that attempts to systematically eliminate waste.*

Some authors have characterized JIT in terms of the so-called 'five zeros' as illustrated in Figure 20.1. JIT is often thought of as a stock reduction technique and there is no doubt that Japanese companies often operate with low levels of inventory by western standards. The objectives of JIT, however, are much more far reaching. JIT can be considered to have a number of elements:

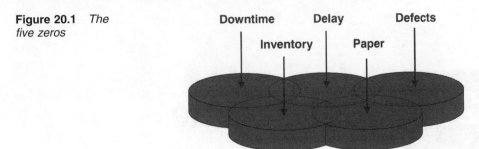

Figure 20.1 *The five zeros*

Set-up time reduction. Perhaps the most spectacular example of manufacturing excellence in Japan is set-up time reduction. Many Japanese companies embraced the *single minute exchange of dies* (SMED) methodology. During the late 1970s, many western companies were astonished that changeovers that took them hours could be completed in Japan in minutes, or in some cases seconds.

Total productive maintenance (TPM). Traditionally, maintenance in the west has focused on corrective maintenance: i.e. reacting to machine breakdowns rather than preventing failure.

Total quality management (TQM). Japanese companies have enthusiastically adopted the ideas of W.E. Deming. Statistical process control (SPC), for example, is employed widely in Japan.

Use of appropriate plant layout. Many Japanese manufacturers have employed group technology (GT) in planning the layout of their factories.

Supply chain management. Japanese companies have adopted a less adversarial relationship with suppliers and have emphasized waste elimination across the supply chain.

Use of 'pull' control systems. In the 1970s, Toyota pioneered the use of 'pull' control techniques, particularly the use of Kanban.

Overall, JIT can be seen as an *integrated* approach to manufacturing. It emphasizes continual improvement. In the areas of human resource management, JIT aims to involve all members of the organization in the improvement process.

20.3 Kanban

'Kanban' is Japanese for card. It is a technique for building the control mechanism into the manufacturing system itself. This can be contrasted with SFC where control comes from outside.

There are a number of kanban techniques that can be employed. The simplest form is so-called *kanban squares*. Here squares are painted

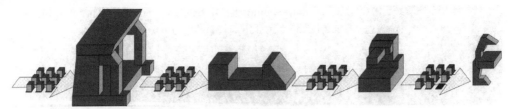

Figure 20.2 *Kanban squares*

between the workcentres for each item in production. When a downstream kanban square is empty, this is the signal for material to be processed. The great virtue of this technique is simplicity as shown in Figure 20.2.

There are several other methods of applying 'pull' control in addition to kanban squares (e.g. kanban cards are also widely applied). All of the techniques operate on the same principle, however: control is an integral part of the manufacturing system. In push systems (like MRPII), control is a separate entity.

Two card kanban. This is a more sophisticated approach. This technique depends on using production (P) and withdrawal (W) kanban cards and standard containers. The kanbans contain simple information relating to the parts to be produced. The cycle can be represented as a five-stage procedure (see Figures 20.3 to 20.7).

1 An empty container arrives at an output queue from a downstream workcentre. Attached to this container is a W kanban.

Figure 20.3 *Two card kanban (stage 1)*

2 The W kanban authorizes material withdrawal and a full container is sent to the workcentre downstream. The P kanban attached authorizes the workcentre to manufacture product.

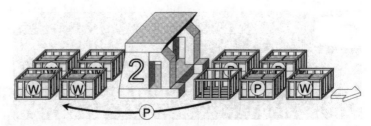

Figure 20.4 *Two card kanban (stage 2)*

3 The workcentre completes a batch of product. The P kanban is attached to the now full container in the output queue.

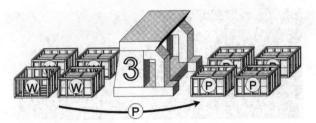

Figure 20.5 *Two card kanban (stage 3)*

4 The empty container in the input queue is sent to the upstream workcentre along with its attached W kanban.

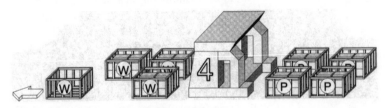

Figure 20.6 *Two card kanban (stage 4)*

5 The upstream workcentre is authorized to send a full container.

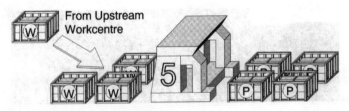

Figure 20.7 *Two card kanban (stage 5)*

Single card kanban. This technique is simpler in that only withdrawal kanbans are used. The five stage cycle is as shown in Figure 20.8. The process is similar to two card kanban except production is controlled externally. This approach can be considered to be a hybrid push–pull system.

20.3.1 Other kanban techniques

An example of a kanban card is shown in Figure 20.9. There are a number of other techniques available for implementing a pull system

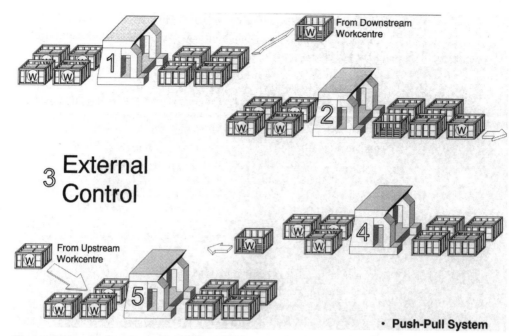

From Downstream Workcentre

₃ External Control

From Upstream Workcentre

• **Push-Pull System**

Figure 20.8   *Single card kanban system*

Figure 20.9
Example of a kanban card

Part Number:	DP1035
Description:	Disk Brake Pad
Box Capacity:	20
Box Type:	A
Issue Level:	3
From: Pressing Cells 1-6	To: Heat Treat

Example Card

Figure 20.10
Other kanban techniques

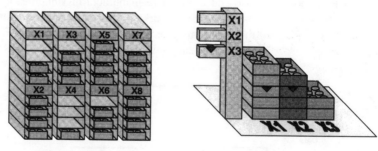

Kanban Racks   **Signal Kanbans**

(see Figure 20.10). Some companies have used so-called signal kanbans. Here, when inventory reaches a predetermined level, a kanban is hung on a signal post where it is highly visible. In other cases, where there is restricted space, racks have been applied (kanban racks). Finally, some companies have developed methods for transmitting the production signal to a remote location. This can be accomplished in a variety of ways; lights, semaphore and even rolling coloured golf balls down transparent tubing.

20.3.2 Calculation of number of kanbans

Kanban is a variant of re-order point (ROP) control. There is no fundamental difference between the two techniques. Rather, they are distinguished by the way in which they are applied. The size of a kanban can be calculated using the following formula:

$$N = \frac{(A_d L_t + S_s)}{C} \tag{20.1}$$

where N = number of kanbans, A_d = average demand over unit time, L_t = lead time, S_s = safety stock and C = container size.

All of the parameters above need to be stated in consistent units. Typically, S_s is stated in terms of a percentage of the average demand during the lead time. Sometimes, however, S_s is calculated using ROP methods by taking into account variation in demand.

20.4 ERP extensions to support kanban

Key:
STK = Stock
RM = Raw Material Stores
FGI = Finished Goods Inventory

The JIT philosophy has a direct influence on the design of Enterprise Resource Planning (ERP) software packages. It is common for ERP packages to include modules for rate levelling. Simplified methods for exchanging information between customer and suppliers have also been developed. These systems often utilize Electronic Data Interchange (EDI), especially in the automotive sector. Many software houses have also added repetitive manufacturing modules to their packages. These modules permit works orders to be handled more easily. In particular, these systems often allow backflushing (see Figure 20.11). This is a technique that allows all of the inventory transactions associated with a

Figure 20.11
Backflushing

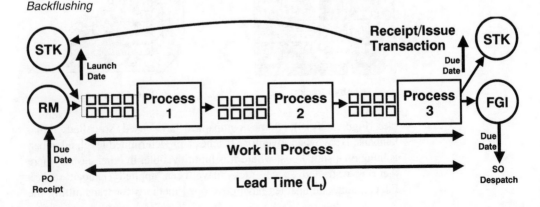

works order to be completed at the point of receipt. This uses the bill of material to calculate the consumption of components for a particular works order, thus eliminating clerical effort. In complex assembled items that are manufactured repetitively (e.g. cars and televisions), this saving can be very significant. It should be noted, however, that backflushing is only appropriate where manufacturing lead times are very short.

20.5 Prerequisites for kanban

Kanban is a very simple yet efficient means of control. There are, however, a number of prerequisites before kanban can be employed as the sole means of control.

Repetitive manufacture. Clearly, kanban cannot be used in an engineer to order environment.

Machine layout. Kanban cannot be used in a functional layout.

Small lot sizes. In practice, for kanban to be successful it is essential that lot sizes are small. As has been seen earlier, this implies short changeover times.

Reliability. Because of the high degree of interdependence between processes, it is essential that there are few disturbances to flow. This is why total productive maintenance (TPM) and total quality management (TQM) are considered to be so fundamental to JIT.

Multi-skilled environment. This is not strictly a prerequisite, but in practice the ability to move labour between processes makes it easier to manage capacity.

Stable demand. Kanban cannot respond to highly fluctuating schedules.

20.6 Influence of JIT

Even in environments where Kanban is not (or cannot) be employed, JIT has been very influential. Early application of MRPII was undertaken in isolation. Fundamental business issues would be addressed by use of safety stock and/or sophisticated planning techniques. The JIT philosophy is to attack the problems themselves. This is simply a recognition of reality: not all problems can be overcome by application of planning and control techniques. Many companies now apply the JIT philosophy of simplification, elimination of waste and continual improvement even if kanban is not the most appropriate scheduling technique in their business.

The JIT philosophy has a direct influence on the design of MRPII software packages. Many software houses have added repetitive manufacturing modules to their packages. These modules permit works orders to be handled more easily. In particular, these systems often allow backflushing. This is a technique that allows all of the inventory transactions associated with a works order to be completed at the point of receipt. This uses the bill of material to calculate the consumption of components for a particular works order, thus eliminating clerical effort. In complex assembled items that are manufactured repetitively (e.g. cars and televisions), this saving can be very significant. It should be noted, however, that backflushing is only appropriate where manufacturing lead times are very short.

Because of the emphasis on simplification and involvement of people at all levels, JIT has led to the adoption of decentralized control strategies. Traditionally, company management structures have been centralized and run on a functional basis. In a JIT environment, many functions (such as planning and control) are carried out by production personnel.

Finally, JIT emphasizes short lead times and small lot sizes and this has enormous benefits for MRP. Consider the example product in Figure 20.12. The MRP table in Figure 20.13 shows that supply and demand

Figure 20.12
Example product

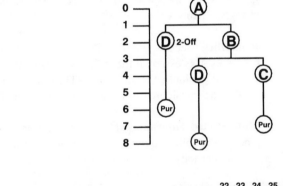

Figure 20.13
Effect of instability
(1)

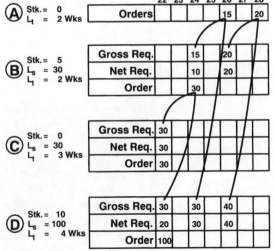

Figure 20.14
Effect of instability (2)

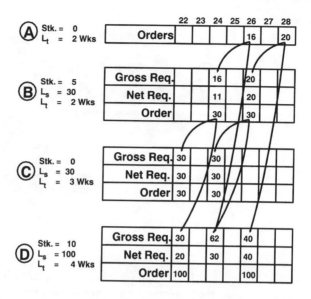

In balance. In Figure 20.14, however, a single additional requirement has been introduced. Because of the lot sizes and lead times in place, this causes large overdue demands. If a closed-loop MRP system was in place, this would create an avalanche of action messages. Often, this instability is blamed on the MRP logic. In reality, the fundamental issue is that the business is insufficiently responsive and the system is merely a reflection of this fact.

Traditionally, in the west, problems have been seen as something to be avoided. In factories, problems such as poor quality, unreliability of machines, etc., have been addressed by the use of safety stocks. Sophisticated techniques have been developed to determine the appropriate level of these stocks. In addition to this, companies have invested in complex computer systems for planning and control that help to avoid such problems in manufacturing.

This approach is sometimes likened to a ship negotiating a river (the so-called *river and rocks analogy*). The western approach is to increase the level of water to allow the ship to pass (in Figure 20.15). Where some obstacles are too large to be covered, radar is used to avoid the problems.

Figure 20.15 *The Classical Approach*

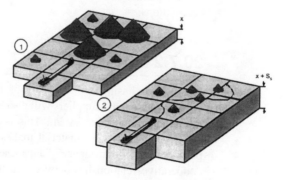

Figure 20.16 *The JIT approach*

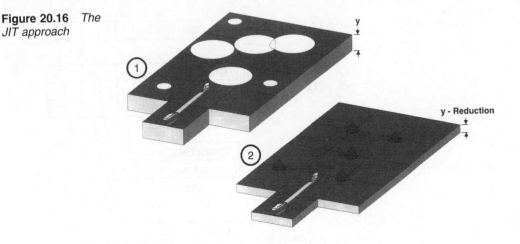

The JIT approach is fundamentally different. Here, the root causes of the problems are attacked using a variety of techniques (e.g. TQM, TPM and SMED). Then, stock levels are reduced and this reveals other problems which themselves are attacked as depicted in Figure 20.16. This process is repeated indefinitely, leading to continuous improvement (or kaizan).

20.7 JIT and purchasing

20.7.1 Traditional purchasing strategies

Traditionally, purchasing has been regarded as separate from the overall manufacturing system. Production planning control would generate requisitions and these would be converted by purchasing into orders. One reason for this was that labour was normally the most significant factor influencing cost. Typically, the majority of purchasing effort was spent on expediting overdue orders. This is often due to an absence of overall planning systems to realistically determine purchase requirements. Purchasing normally focused on price and the relationship with suppliers was adversarial. Large numbers of suppliers were perceived as a good thing. This allowed purchasing to apply pressure during price negotiations.

20.7.2 Influence of JIT

Ideally, purchasing will be part of the manufacturing management system. Indeed, the suppliers will also be considered an integral element in the manufacturing system. This is particularly important in manufacturing today where material makes up a significant part of overall product cost. The influence of Japanese manufacturers has led to a more cooperative relationship between customer and supplier. Increasingly,

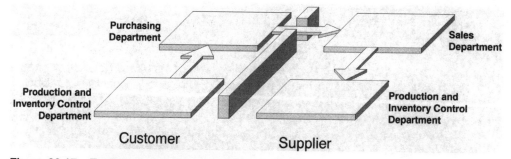

Figure 20.17 *Traditional customer – supplier relationship*

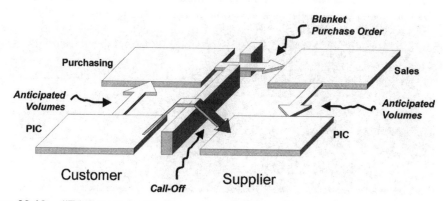

Figure 20.18 *JIT influenced customer – supplier relationship*

price is only one factor in choosing a supplier. Perhaps most important is the trend toward single sourcing and the emphasis on minimizing cost across the entire supply chain.*

20.7.3 Customer–supplier communication

One recent development in many companies is that the production and inventory control (PIC) departments of customer and supplier communicate directly (as illustrated in Figures 20.17 and 20.18). This is essential as lead times are reduced. This leads to a change in the role of purchasing: increasingly, they are concerned with strategic objectives such as rationalizing the supplier base. Managing the supply/demand relationship, which is essentially a tactical function, is now part of the mainstream manufacturing management system.

* Further information on purchasing can be found in *Business Skills for Engineers and Technologists* (ISBN 0 7506 5210 1), part of the IIE Text Book Series.

20.8 Summary

While there is no generally accepted definition of JIT, it is best thought of as an approach that attempts to eliminate waste. It is a decentralized approach that means people at all levels in the organization are involved in decision making. While many companies have embraced these ideas, in some cases the application has been selective. JIT, for example, emphasizes the involvement of people at all levels. Some companies have misinterpreted the JIT philosophy and have simply forced staff to undertake a wider range of tasks. Similarly, JIT encourages customer–supplier partnership. There are examples, however, of suppliers being forced to supply 'JIT' with no assistance or support. It is also now being recognized that traditional techniques also have their virtues and can also be employed.

There is no doubt, however, that JIT is the most influential idea in industry in the west over the last 20 years. It is a philosophy that requires changes not only to plant and its layout, but also attitudes throughout organizations. Where JIT principles have been applied enthusiastically and comprehensively, great benefits have resulted.

21 Quality

21.1 Overview

Quality is an essential element of the just-in-time (JIT) philosophy. Like JIT, there is no universal definition, but perhaps the best was proposed by Crosby:

'Quality is conformance to requirements'.

The term 'quality assurance' dates back to the 1920s and was coined by a small group, including the statistician Walter Sheward working at the Bell Telephone Laboratories. They developed a number of techniques including statistical process control (SPC). These techniques were developed by W. Edwards Deming between 1930 and 1950. Most significantly, he visited Japan in the 1950s and gave a series of lectures on quality. The techniques he taught were adopted enthusiastically and eventually developed into total quality management (TQM). Application of quality techniques has now extended beyond manufacturing. It is now employed in organizations as diverse as hotels, retailers and software houses.

This chapter will examine the ways in which quality has impacted on industry in general.

21.2 Statistical process control

21.2.1 SPC theory

SPC was originally devised for use in manufacturing companies, though it has now been applied in a range of organizations. Fundamental to SPC is the idea that deviations from a mean value can arise in two ways:

Random variation. Most manufacturing processes are *stochastic* – that is to say, they are subject to random variation to a greater or lesser extent. They cannot, therefore, be defined by a single value and a probability distribution is required.

Special causes. Here some event has occurred that has led to a deviation.

Sheward (and later Deming) realized that these two types of deviation needed to be treated differently. In the case of a special cause, then a corrective adjustment would need to be made. If, however, the deviation was the result of random variation, then the only solution would be to improve the process capability (i.e. reduce the range of process deviation). The question naturally arises, how can these types of deviation be distinguished? The solution is to consider the problem statistically.

Consider Figure 21.1. This shows the diameter of a particular component measured over a period of time. On some occasions, the component is outside limits, on others inside. Notice, however, that there is no particular trend in the data. In this case, it is pointless making an adjustment.

In Figure 21.2, note that the component is within tolerance on each occasion. It might be inferred therefore that there are no problems and no adjustment is required. There is clearly a trend, however, and if no action is taken, future components will be outside limits. In both of these cases, the process is *out of control*.

Figure 21.1   *Out of control process (1)*

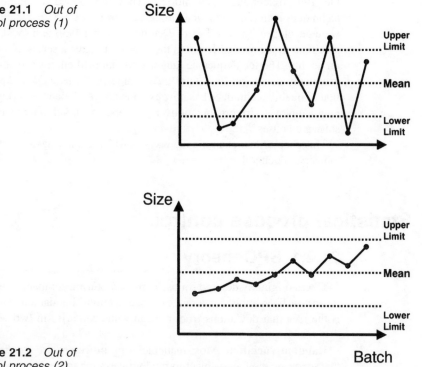

Figure 21.2   *Out of control process (2)*

Figure 21.3 *In control process*

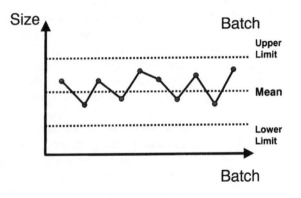

A process is said to be *in control* (as shown in Figure 21.3) if:

- Most points are close to the mean and few near to the limits.
- Comparable numbers of points are above and below the mean.
- No pattern in the data exists (e.g., no trend up or down).

When processes are out of control it is necessary to *mass inspect*, as it is impossible to predict whether a particular component will be outside limits. The first step in improving any process therefore (whether manufacturing or otherwise) is to bring it under control. Once processes are in control, it is only necessary to *sample inspect*.

It might be thought that SPC is purely a technical approach to improving quality. As Deming has pointed out, however, it has implications for work practice and the way in which people are managed. If a process is out of control, there is no point in penalizing an operator for scrap or providing incentives for the converse. From these principles, Deming produced his so-called *14 points for management*.

1 Create a constancy of purpose toward product improvement to achieve long-term organizational goals.
2 Adopt a philosophy of preventing poor-quality products instead of acceptable levels of poor quality as necessary to compete internationally.
3 Eliminate the need for inspection to achieve quality by relying instead on statistical quality control to improve product and process design.
4 Select a few suppliers or vendors based on quality commitment rather than competitive prices.
5 Constantly improve the production process by focusing on the two primary sources of quality problems, the system and workers, thus increasing productivity and reducing costs.
6 Institute worker training that focuses on the prevention of quality problems and the use of statistical quality control techniques.
7 Instill leadership among supervisors to help workers perform better.

8 Encourage employee involvement by eliminating the fear of reprisal for asking questions or identifying quality problems.

9 Eliminate barriers between departments, and promote cooperation and a team approach for working together.

10 Eliminate slogans and numerical targets that urge workers to achieve higher performance levels without first showing them how to do it.

11 Eliminate numerical quotas that employees attempt to meet at any cost without regard for quality.

12 Enhance worker pride, artisanry (*or craftsmanship*) and self-esteem by improving supervision and the production process so that workers can perform to their capabilities.

13 Institute vigorous education and training programmes in methods of quality improvement throughout the organization, from top management down, so that continuous improvement can occur.

14 Develop a commitment from top management to implement the previous 13 points.

Deming makes the point that the purpose of quality management is to check the process, not the product (customers will check the product). Furthermore, he also states that quality cannot be inspected into a product. SPC shows that simply employing more inspectors cannot improve quality.

21.3 Quality costs

21.3.1 Types of cost

There are two types of cost associated with quality. First, the cost of conformance (COC); the cost of ensuring goods or services conform to specification. Second the cost of non-conformance (CONC). These costs can be further subdivided as shown in Table 21.1.

21.3.2 Quality – cost trade-off

Traditionally, manufacturers took a view that some defects were acceptable (the so-called, 'acceptable quality level' or AQL). The reason for this was because it would cost too much to eliminate all defects (see Figure 21.4). In recent years, however, this belief has been challenged. This is for two reasons:

- COC is lower than previously thought. Many of the ways in which quality can be improved are simple. For example, if processes are designed such that it is difficult for operators to make mistakes, this can greatly reduce defects. In Japan, this approach is called *Poke-Yoke* or fool proofing.
- CONC is much higher than previously thought. The impact of poor quality on customers is very high (especially today). Defects will almost certainly cause customers to defect to other suppliers.

Table 21.1   Quality costs

Conformance		Non-conformance	
Prevention	*Appraisal*	*Internal failure*	*External failure*
Quality planning: developing quality systems.	**Inspection:** testing of products or services throughout processes.	**Scrap:** products or WIP that need to be discarded.	**Complaints:** the effort required to respond to customers.
Design: building quality into the design of product or service.	**Equipment:** running and depreciation of test equipment.	**Rework:** correction of problems to restore to conformance.	**Returns:** product replacement and administration.
Process: cost expended on ensuring process conformance.	**Operators:** time needed to gather data and assess quality.	**Investigation:** the effort needed to find cause of problem.	**Warranty:** complying with stated obligations (possibly statutory).
Training: operators, supervision and management.		**Downtime:** loss of resources/time while resolving problems.	**Liability:** involvement in litigation, possibly involving injury.
Information: systems for data collection and analysis.		**Downgrading:** the penalty of having to sell 'seconds'.	**Lost sales:** customers may defect to other suppliers.

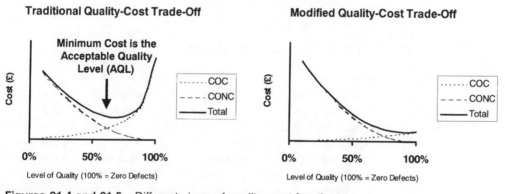

Figures 21.4 and 21.5   *Different views of quality–cost function*

Compared with Figure 21.4, it is more realistic to draw the quality–cost trade-off curve as shown in Figure 21.5.

This has led to the concept of zero defects (ZD). Here, there is no AQL and the goal is to eliminate all defects. In some world-class manufacturing companies, defects are measured in parts per million (PPM). Note that ZD is now widely employed in service organizations.

21.4 Total quality management (TQM)

21.4.1 Overview

The culmination of many of the ideas discussed previously is TQM. The basic idea behind TQM is that all employees in the company are responsible for quality, with senior management taking the lead. This is in contrast to the traditional approach where quality was the responsibility of a specialist department. Furthermore, TQM aims for zero defects. Because elimination of all defects is impossible, some authors have remarked that TQM is a journey, not a destination. Thus continuous improvement is a crucial element of the concept. The key elements of TQM are:

- Customers define quality. Customer requirements are paramount.
- Senior management provides quality leadership.
- Quality is a strategic issue. Quality is the key focus for strategic planning.
- Quality is the responsibility of all.
- Continuous quality improvement is key. All company functions must focus on this to meet objectives.
- Cooperative effort is crucial. All employees must be involved in solving problems.
- Statistical methods are key. These are the basis of problem solving and continuous improvement.
- Education/training are essential for all personal. This is a prerequisite for continuous improvement.

21.4.2 Human factors

Because TQM rejects the notion that quality is the responsibility of specialists, human factors are extremely important. Employee involvement (EI) is crucial to the process. Note that this is in contrast to the ideas of scientific management, where individual workers were expected simply to follow instructions. One concept that many companies have employed successfully is quality circles. This is a small, voluntary group of employees (in Japan, personnel typically receive no extra pay for this activity), who work together to solve problems. The operation of quality circles is summarized in Figure 21.6.

Quality circles have been extremely successful in Japan, though less so in the west. It is interesting to note that in the USA, quality circles have been most successful in non-manufacturing companies.

21.5 Certification

In the 1970s, it was recognized in Britain that standards were required for quality. This led the British Standards Institute (BSI) to formulate

Figure 21.6
Quality circles

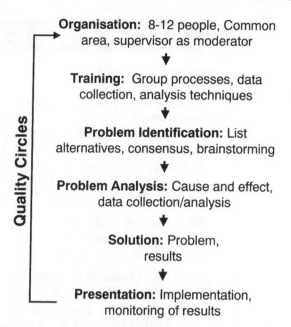

Organisation: 8-12 people, Common area, supervisor as moderator

↓

Training: Group processes, data collection, analysis techniques

↓

Problem Identification: List alternatives, consensus, brainstorming

↓

Problem Analysis: Cause and effect, data collection/analysis

↓

Solution: Problem, results

↓

Presentation: Implementation, monitoring of results

BS5750: a standard whereby companies could be certified for procedural integrity. This standard was adopted by the International Standards Organization (ISO). The resulting standard was denoted ISO9000. Companies can pay for independent auditors to check their documentation and if this is adequate, they receive certification. It is now common in the UK for customers to refuse to do business with suppliers who are not ISO9000 certified.

The ISO9000 standard does not itself relate to quality; it simply outlines a systematic approach to documentation. It is for this reason that the ISO9000 standards are relatively short. Thus, there is a danger that companies who have been certified to ISO9000 believe they have good quality. This is not true; ISO9000 accreditation simply means that the company concerned has a comprehensive and consistent set of documents outlining business practices. It does not mean these practices are optimal or even effective.*

21.6 Summary

Quality has undergone a revolution in recent years and TQM has gained widespread acceptance. It should be recognized that while TQM is a management philosophy, its roots are in hard statistics. In recent years, certification has become common and ISO9000 is often a prerequisite for doing business. Finally, all of these quality concepts have been developed from a manufacturing base. They are now employed, however, across virtually all enterprises.

* Further information on quality management can be found in *Business Skills for Engineers and Technologists* (ISBN 0 7508 5210 1), part of the IIE Text Book Series.

21.7 Case study

21.7.1 ComNet

ComNet are a network support company with 200 employees. The company installs PC networks of varying sizes for local customers. The company also has a number of highly lucrative support contracts. The company has expanded rapidly and is profitable, though the management team are vaguely concerned things are going wrong. The following is a report of a senior management team meeting.

MD = Managing Director
FD = Finance Director
OM = Operations Manager

MD   I am extremely annoyed. I was telephoned yesterday by an irate customer saying that some wiring hadn't been correctly installed. As a result, a cable went astray and someone tripped and brought the network down. What is going on? It looks like a basic error to me.

OM   The guy was a trainee – they tend to be careless. Their work needs to be checked.

MD   The customer told me one of our supervisors HAD been on site and HAD checked things out.

OM   The supervisor has only been recently promoted. He must have taken his eye off the ball.

MD   What we need to do is to check each piece of work twice – maybe we need a system of signing off these jobs by all concerned. Three signatures should ensure the job's done right.

FD   The real solution to this problem is a rapid response team. When we've made mistakes, we need to get a team of our best people out on site ASAP to minimize customer complaints. Trying to eliminate all mistakes would cost a fortune and is probably impossible.

MD   There are too many complaints overall. Our biggest problem seems to be network cards.

OM   Two issues there. First, it's the same thing as before, trainees not installing cards correctly – basic mistakes. Second and more importantly, the cards themselves are causing problems. Maybe we should change supplier?

FD   I'd be reluctant to do that. The supplier gives us an excellent price based on the large volumes we buy. Incidentally, we need to increase the size of the stores. We're having to store some of the cards on the floor outside – they're actually violating fire regulations.

OM   OK, but it's going to take some time. I'm rushed off my feet sorting out these network card problems on site. There's only 24 hours in a day.

FD   We could try an incentive scheme. Each month we could put the names of the installation team with the fewest network problems up on the notice board. We could give them a 'best team of the month' bonus too.

OM   In my opinion, we should also put the 'worst team of the month' up on the notice board.

MD   I'm also a little concerned about the situation at the bank. We're getting new business, but some customers are withholding payments until we rectify problems. The size of the overdraft is getting too large – the bank doesn't like it. The cost of those network cards didn't help.

FD   It'll payoff in the long term – I got an excellent volume discount. We won't need to buy any more for maybe six months or so.

MD   At least that's good news.

FD   There's another advantage of having the high stocks – we have enough to send out extra cards with the installation teams. That way, if a card falls over, we can replace it without having to drive back to base. We've been wasting far too much money on driving back and forth.

OM   Incidentally, one of the trainees thought we should build a simple rig to test cards before being taken out on site and fitted.

MD   Which trainee was that?

OM   The one who caused the loose cable problem.

MD   Tell him I'll consider his suggestion on the day he's learned to fasten down a cable. He should spend less time thinking and more time doing his job.

OM   I've also been asked by the trainees again about courses . . .

MD   I'm fed-up of hearing about courses – training is expensive and right now, we're short of money. A lot of our problems are the fault of the trainees and now they're asking for courses! Which part of 'no' don't they understand?

21.7.2 Analysis

ComNet certainly have not adopted the principles of TQM. Even at the most basic level, management could be improved. Consider the following points:

1 Management should have listened to the reasonable suggestion for a test rig.
2 The company has too many network cards – why buy six months of stock?
3 Are the network card problems the fault of the suppliers or the poor storage. This could be checked. Even if the cards are a problem, it is going to take a long time to work through the inventory. What data do ComNet have to convince the supplier to take back the whole batch of cards?

4 An incentive scheme would seem to be useless, since the biggest problem seems to be the network cards themselves. How could being the best or worst team be anything other than a lottery?

5 Will further checking help – one supervisor didn't make any difference, why should two?

6 Why not sample check the work of trainees? Those who are struggling should be given help. This would also save a lot of time.

7 Should the request for training be dismissed so lightly? If the trainees are really a source of problems, doesn't this imply they need more training, not less?

8 Is a rapid response team a good idea – wouldn't getting it right first time be better? Surely the customer would think so.

9 The cost of non-conformance is clearly very high – the operations manager is rushed off his feet. Some effort should be diverted into prevention.

22 Plant management

22.1 Overview

Plant management has been one of the less glamorous functions in industry. Yet failure of plant is inevitably disruptive and today, more than ever, can have a serious negative effect on business performance. Failure rates over time have traditionally been described by the graph in Figure 22.1.

In the short term, there will be a relatively large failure rate due to problems present at start-up. This will be followed by a long period of stability. Finally, as the plant wears, failure rates increase dramatically. Failure rate may be expressed in terms of time or usage as defined by equations (22.1) and (22.2).

$$\text{Failure rate} = \frac{\text{Number of failures}}{\text{Total quantity produced}} \times 100\% \text{ (failures per unit)}$$

$$(22.1)$$

or

$$\text{Failure rate} = \frac{\text{Number of failures}}{\text{Operating time}} \times 100\% \quad \text{(failures per hour)}$$

$$(22.2)$$

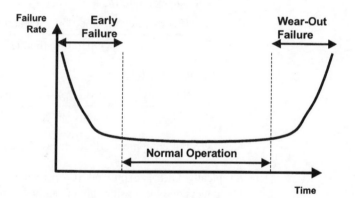

Figure 22.1
Classical failure curve

In practice, the shape of the failure curve will vary depending on the nature of the plant. All plant, however, is subject to failure and organizations need to provide suitable infrastructure to address this issue. The discipline of plant management is referred to as terotechnology. This is defined as follows:

> *A combination of management, financial, engineering, building and other practices applied to physical assets in pursuit of economic life-cycle costs.*

As equipment has become more sophisticated and costly, maintenance and plant management have increased in importance. At a purely financial level, the greater the cost of plant, the greater the importance of ensuring it provides a good return on investment (ROI).

Perhaps the most significant factor, however, in the growing recognition of the importance of plant management is the high priority given to maintenance by Japanese companies. During the 1960s and 1970s in Japan, total productive maintenance (TPM) was developed. TPM is now widely practised in world-class organizations (this will be discussed later in this chapter).

This chapter will review some of the basic techniques for managing manufacturing plant.

22.2 Overall equipment effectiveness

As discussed in 3.5, engineers need to justify expenditure on plant and equipment. Once plant has been purchased, however, it is essential that it operates effectively. A key measure that can be applied to manufacturing plant is overall equipment effectiveness (OEE). This is defined below in equations (22.3) to (22.6):

$$\text{Availability } (A_v) = \frac{(\text{Available time } (T_a) - \text{Downtime } (T_d))}{\text{Available time } (T_a)} \tag{22.3}$$

$$\text{Performance rate } (R_p) = \frac{\text{Theortical cycle time } (T_c) \times \text{Quantity processed } (Q_p)}{\text{Operating time } (T_o)} \tag{22.4}$$

$$\text{Quality rate } (Q_R) = \frac{(\text{Quantity processed } (Q_p) - \text{Quantity scrapped } (Q_s))}{\text{Quantity processed } (Q_p)} \tag{22.5}$$

$$\text{Overall equipment effectiveness (OEE)} = A_v \, R_p \, Q_R \tag{22.6}$$

Typically, OEE is expressed as a percentage. The equations above can be understood by considering a machine running for two hours. If there is one hour of downtime, there will be only 50% of the time available for useful production (i.e. availability = 50%). Downtime can arise for a variety of reasons. There may be a physical reason (e.g. machine set-up or breakdown). Alternatively, downtime may arise because of a logistical deficiency (e.g. an operator is not available). Of this 50%, if the machine is running below the theoretical operating speed this will further reduce effectiveness. Thus if a machine operates for one hour and in this time produces 45 products, which nominally should be produced with a cycle time of 1 minute per unit, this will give a performance rate of 75%. Finally, if 9 of the 45 products are scrapped, this will yield a quality rate of 80%. Thus OEE = 50% × 75% × 80% = 30%. Thus in two hours, only 30% of the theoretical production of 120 units (30% × 120 = 36) is achieved.

22.3 Plant maintenance principles

22.3.1 Maintenance strategies

Fundamentally, there are three different maintenance strategies:

- **Breakdown maintenance.** Here, maintenance is undertaken in response to a problem that has been detected during use of equipment.
- **Preventive maintenance.** Here, maintenance is undertaken at routine intervals. These intervals may be determined by elapsed time or time of in-service use of equipment.
- **Predictive maintenance.** In this case, equipment is routinely inspected (e.g. vibration levels may be monitored). Based on this inspection, maintenance activities may be triggered.

22.3.2 Traditional approach to maintenance

The traditional approach to maintenance was to trade off the benefits of undertaking preventive maintenance against its cost. The premise was that there was an optimum level of preventive maintenance as shown in Figure 22.2. As more preventive maintenance is undertaken, the cost of breakdowns will fall. The cost of carrying out the preventive maintenance will of course increase. In this model, a certain level of breakdowns is acceptable.

The problem with this model is that the cost of breakdown is difficult to quantify. Traditionally, the costs have been defined in terms of the number of hours of lost production. In practice, however, the real cost is much higher. For example, a breakdown might cause an order to be supplied late causing dissatisfaction and potentially, the loss of a customer to a competitor. A more realistic representation of the situation is shown in Figure 22.3.

Figure 22.2
Traditional maintenance model

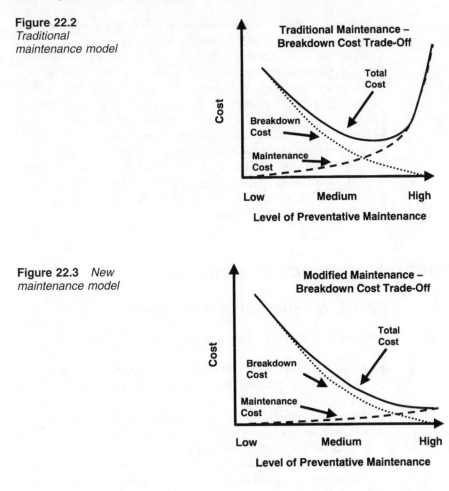

Figure 22.3 *New maintenance model*

In practice, breakdowns are more serious in just-in-time (JIT) as large inventories of products are not available to provide a buffer against unreliable plant. Many Japanese companies have a target of zero breakdowns.

22.3.3 Use of CMMS

Computer maintenance management systems (CMMS) are in common industrial use. These are computer packages that automate some of administration associated with maintenance activities. CMMS packages can be highly sophisticated and complex. The basic structure of such systems, however, is relatively simple (see Figure 22.4). CMMS packages consist of a number of modules as described below.

- **Plant register.** The module allows all plant items to be logged.
- **Preventive maintenance schedules.** This module allows maintenance schedules to be defined. These can be linked to particular plant items.

Figure 22.4 *Basic CMMS package structure*

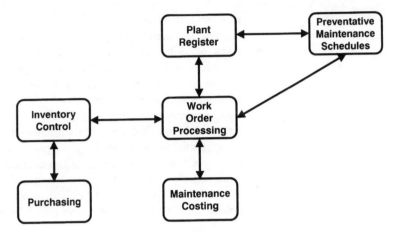

- **Work order processing.** This module allows work orders to be created. Work orders can be created manually in response to breakdowns or automatically as a result of a preventive maintenance schedule.
- **Inventory control.** Spare parts may need to be issued to a particular works order. Inventory control keeps track of stock levels of spares. This module will also recommend that parts are ordered when stock levels fall below a defined reorder point.
- **Purchasing.** This module allows spare parts to be ordered and received.
- **Maintenance costing.** This module allows maintenance costs (labour and spares) to be collated and attributed to particular machines or groups of machines.

It should be recognized that implementing a CMMS package in itself does not improve maintenance; it simply automates administrative activities. Many companies have experienced difficulties when implementing CMMS packages. There are a number of causes for this, including:

- **Lack of commitment.** The benefits of avoiding breakdowns are well recognized by most companies. These benefits, however, lag the effort required to apply preventive or predictive techniques (i.e. the number of breakdowns reduces only sometime after predictive/preventive techniques are applied). Because of this, some companies are overwhelmed by breakdowns and never get around to changing their operating methods. One way to alleviate this problem is to implement the new techniques in stages, spreading the effort over a longer period. This also has the effect of demonstrating the benefits of modern maintenance methods.
- **A breakdown culture.** This is related to the point above. The personnel in many maintenance departments are experienced in breakdown maintenance. They are not accustomed to working on machines that apparently are working perfectly well. A further

problem is that production may be unwilling to release machines for maintenance activities (unless forced to do so by breakdowns).

- **Lack of skills.** The skills required for preventive or predictive maintenance are more analytical in nature than those traditionally needed for handling breakdowns. Unless preventive maintenance schedules are continually reviewed, based on the performance of the equipment, the full benefits of the approach cannot be obtained.
- **Excessively detailed maintenance schedules.** A common problem when implementing CMMS packages is that preventive maintenance schedules are defined at too detailed a level. Each time a work order is created for a preventive maintenance task it needs to be printed and subsequently closed. Most successful companies delegate routine tasks to operators (see 22.4 below).

22.4 Total productive maintenance (TPM)

22.4.1 TPM overview

Total productive maintenance (TPM) is a vast subject and can only be touched upon in this book. It is an integral part of the JIT philosophy and is strongly related to total quality management (TQM). The traditional western approach is that maintenance is the responsibility of a specialist function within the organization. The TPM philosophy is that maintenance is the responsibility of the whole company, from senior management to machine operators. TPM promotes the use of preventive and predictive maintenance techniques, but also stresses the importance of the involvement of machine operators. It also emphasizes the importance of applying techniques such as FMEA to design out potential problems with plant. Overall, TPM requires an enormous cultural change for an organization. Two techniques employed to support TPM are described below.

22.4.2 5S approach

Crucial to TPM is order and tidiness in the work environment. TPM promotes the use of so-called 5S activities. The term 5S is derived from the five Japanese words listed below:

- **Seiri (sort).** Separating that which is necessary from that which is not and dispensing with the latter.
- **Seiton (straighten).** Arranging and identifying things for ease of use.
- **Seiso (sweep and clean).** Cleaning and tidying the work area.
- **Seiketsu (standardize).** Defining routine procedures to maintain the above processes.
- **Shitsuke (sustain).** Ensuring an appropriate culture is in place to support the other four rules.

22.4.3 SMED approach

Set-up time is defined as the time taken to change a machine from producing one type of product to another. Long set-up times reduce OEE, but more important, they force companies to manufacture in large batches or lots. Small lot sizes are also important in ensuring that customers can be supplied in the quantities they require without the necessity of holding large stocks. If lot size is reduced (without first reducing set-up time), however, this can have negative consequences. In particular, small lot sizes will mean too much time is spent setting up and in effect starve the organization of capacity. For this reason, many organizations are engaged in set-up time reduction exercises.

The best-known approach for set-up time reduction is single minute exchange of dies (SMED). While this approach was originally devised for die changeover, it can be applied in a wide range of circumstances. The methodology has four steps:

1 **Analyse current changeovers.** It is first necessary to understand how the existing set-up is undertaken. There are a number of techniques for achieving this, but one of the most effective is to use a camcorder.
2 **Separate internal and external operations.** Internal operations are those that can only occur while a machine has stopped. External operations can be carried out while a machine is running.
3 **Convert internal operations to external.** The best-known example of this is pre-heating in the case of die casting operations. In many companies, however, major benefits can be obtained simply by better organization and ensuring all of the necessary tooling and information is available at the time of the set-up.
4 **Streamline all aspects of the set-up.** It is often the case that expensive and sophisticated technology is not required. Many of the best ideas for process improvement originate from operators and are extremely simple. One obvious avenue of attack is to eliminate adjustments: these invariably waste time and often lead to 'first-off' scrap. There is no formula for set-up reduction, but the following are useful guidelines:
   - Be critical of the existing process. Are all of the existing processes required?
   - Be cautious about adding new elements to the process – simplification is the aim.
   - Think laterally about the process – brainstorming is an excellent technique in this area if the right people are involved.

22.5 Plant management – summary

Effective plant management is now recognized as being crucial to the success of any organization. The philosophy of breakdown-oriented maintenance is now generally discredited. Many companies have attempted to adopt TPM principles. Reliability, particularly in a JIT environment, is now recognized as being crucial. Because TPM requires a cultural change, however, adoption has proved a significant challenge in most western organizations.

Exercises – Planning, scheduling and logistics

Exercises – Open-loop control

1 Consider the MRP example below:

Part number: DP1023   Lead time: 2   Lot size: 3000

	SOH	1	2	3	4	5	6
PGR		1000	1000	1000	1000	1000	1000
Sch Rec			3000				
PAB	1950						
POR							

Part number: PL2349   Lead time: 2   Lot size: 6000

	SOH	1	2	3	4	5	6
PGR							
Sch Rec							
PAB	500						
POR							

Key: SOH – Stock on Hand   PGR – Projected Gross Requirements
      Sch Rec – Scheduled Receipts   PAB – Projected Available Balance
      POR – Planned Order Release

A single unit of PL2349 is required to manufacture DP1023.
(a) Calculate the projected available balance and planned order releases for DP1023.

(b) By projecting gross requirements from DP1023 to PL2349, calculate the projected available balance and planned order releases for PL2349.

(c) Assuming a closed-loop MRP system is in place, what action messages would be generated if the projected gross requirement for DP1023 in week 1 was increased from 1000 to 3000?

2 A make to stock organization supplies a product with the following attributes:

Lot size	3000 units
Lead time	15 days
Average daily demand	1500 units/day

(a) Assuming that the actual demand can be represented by a normal distribution, calculate a reorder point that will theoretically yield a service level of 50%.

(b) What is the theoretical average stockholding based on the reorder point calculated above?

Exercises – Closed-loop MRP

1 A company makes two products sharing a common raw material. The two bills of material are shown below:

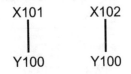

To manufacture one unit of X101 or X102 one unit of Y100 is needed. The company has three customers, A, B and C. Their requirements for the two products over five weeks is as shown below:

		Week				
Product	Customer	11	12	13	14	15
X101	A	10	10	10	10	10
	B	5	5	5	5	5
	C	5	5	5	5	5

X102	A	5	5	5	5	5
	B	5	5	5	5	5
	C	10	10	10	10	10

The data below are associated with the three items (no allowances need to be made for scrap or yield):

Part	Lot size (units)	Lead time (weeks)	Stock on hand (units)
X101	40	2	55
X102	40	2	55
Y100	100	2	5

(a) Both X101 and X102 have scheduled receipts due on week 12 for 40 units. Draw an MRP grid and calculate when planned orders need to be released for the two items.

(b) Assuming no scheduled receipts are in process for Y100, draw an MRP grid and calculate when planned orders need to be released for this item.

(c) What action messages might be generated by a closed-loop material requirements planning (MRP) system?

Exercises – Master production scheduling

1 Outline the reasons why a company using a manufacturing resource planning (MRPII) system might need to employ master production scheduling even if its aggregate workload is relatively stable.

2 Consider the data below which are associated with the master production schedule (MPS) for a particular item:

Part number: FDB9749   Lead time: 2   Lot size: 30

	SOH	1	2	3	4	5	6	7	8
				DTF			PTF		
Independent forecast				10	11	12	12	12	12
Actual demand			10	8	10	20		10	
Proj. available balance	40								
MPS (receipt)					30		30		
Available to promise									

Where:
SOH   Stock on hand
DTF   Demand time fence
PTF   Planning time fence

(a) Why is the forecast in weeks 1 and 2 set to zero?
(b) Calculate the projected available balance from week 1 to 8.
(c) Calculate available to promise (ATP) over the same period (show method).

3 A small company manufactures a range of ride-on lawn mowers. There are a range of different options and a generic bill of material for this product is shown below:

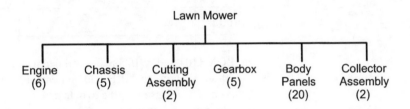

Note that the number in parentheses represents the number of options offered to customers – for example, there are six different engine options. Total sales are around 10 000 products per year. The lead time to assemble the first level subassemblies into finished product is equal to the lead time required by customers. The time to manufacture/procure these subassemblies, however, is much longer.

(a) Assuming there are no other sources of variation except those shown above, how many finished product options are offered to customers?
(b) Explain why it would be inappropriate to hold inventory in the form of finished goods.
(c) Outline a planning system suited to the situation described above.

Exercises – Finite and infinite capacity scheduling

1 A factory has three resources, R1, R2 and R3. Consider the routing below:

	L_s	Op 1			Op 2			Op 3		
		Res	T_p	T_s	Res	T_p	T_s	Res	T_r	T_s
Job 1	10	R1	1.00	4.00	R2	1.50	5.00	R3	2.00	5.00
Job 2	10	R1	1.50	5.00	R3	2.00	10.00	R2	2.00	5.00
Job 3	10	R1	1.50	5.00	R2	1.00	5.00	R3	3.00	5.00

The priority of the jobs is in descending order from the top (job 1 is the most important). Manually schedule the three jobs to finite capacity. What is the makespan?

2 Using Johnson's algorithm, find the optimum schedule to minimize makespan for the example below?

		Machine	
Job	**Time (1)**	**Time (2)**
A1	8	2
A2	2	10
A3	10	9
A4	9	1
A5	7	5
A6	3	6
A7	1	7
A8	11	3
A9	12	4
A10	5	8
A11	6	11
A12	4	12

3 Five jobs are queuing at a workcentre (operation seven). Job D has three operations remaining, job C two and the others one. The factory runs seven days a week on a single shift pattern with eight hours per day. The three jobs below are queuing at a workcentre. The routing table below shows the operations remaining (all times in minutes).

Routing information

		Op 7		Op 8		Op 9		Due date
Job	L_c	T_r	T_c	T_r	T_c	T_r	T_c	
A	100	2	100					11th April
B	100	2	120					10th April
C	100	2	110	2	100			12th April
D	100	2	100	2	100	3	120	13th April
E	100	3	150					11th April

The current date is April 8th. What is the priority sequence using the following rules:

- Earliest due date.
- Shortest processing time.
- C_r.
- L_{Slack}.
- L_{SPRO}.

4 What are the transactions involved in a works order management system? What are the advantages and disadvantages of making these transactions on the shop floor?

5 What are the advantages and disadvantages of using work to lists as a means of controlling a manufacturing system? How can the manufacturing engineer help to simplify the control of the manufacturing system?

6 Explain how a short interval scheduling (SIS) system combines some of the attributes of manufacturing resource planning (MRPII) and FCS. What will be the consequences of introducing a SIS to a company with poor master scheduling disciplines?

Exercises – Just-in-time

1 Manufacturing resource planning (MRPII) is sometimes described as a 'push' system. Just-in-Time (JIT) is usually described as a 'pull' system.

   (i) At a fundamental level, what is the difference between a 'pull' and a 'push' system?
   (ii) The simplest 'pull' system is the kanban squares method. Explain why some companies employ the more complex two card kanban technique.

2. Set-up time improvement and lead-time reduction are often associated with the JIT philosophy. Discuss how such improvements would influence a company using MRPII.

Exercises – Quality management

1 The chief executive of a large manufacturing group has written the following memorandum to all of the factory managers in the company.

From: J. Burton
To: All Plant Managers
Date: 17th January 2004

Subject: **Just-in-Time (JIT) and Total Quality Management (TQM)**

As you are probably aware, JIT and TQM are concepts that have been widely discussed in the press. I feel that we have not adopted these techniques with sufficient vigour. Therefore, I would like a detailed plan from each of you explaining how JIT and TQM will be implemented in your particular factories. I would expect JIT and TQM to be fully implemented by the end of the year so that we can focus once more on cost reduction.

JIT is principally about stock reduction. I would suggest therefore that our suppliers are told that they need to supply product to our plants on demand so that we do not need to hold stocks. It should be made clear to them that unless they agree to these terms, we will move our business to other suppliers. I would also suggest Kanban is implemented by the end of February in all plants and queues between machines cut to a tenth of their current value at this time.

In the area of TQM, it is essential that employees causing quality problems be identified. I propose that each plant produces a monthly list of employees and the percentage of defects they produce. This list should be placed on noticeboards and the worst employee interviewed by his/her manager. If an employee is at the bottom of the list two months in a row, they should receive a formal warning. If an employee is at the bottom of the list for a third month, they should be dismissed.

Finally, I would propose that we implement a suggestion scheme from employees. I believe that the workforce can produce many useful ideas for improving efficiency. We need to focus particularly on those suggestions that can reduce our labour requirements. This is especially important at present, as I would expect you to need to recruit more inspectors as part of your TQM programmes.

I look forward to receiving your plans by the end of the month. I should remind you that in the current climate, these plans should be possible within current budgets.

The memo claims to promote JIT and TQM principles. Do you think the memo actually does this? Give a reasoned argument to support your view.

Exercises – Plant management

1 Briefly outline the three fundamental maintenance strategies.

2 Set-up time improvement and lead-time reduction are often associated with the JIT philosophy. Discuss how such improvements would influence a company using MRPII.

References

Checkland P. (1981) *Systems Thinking, Systems Practice*. John Wiley and Sons, New York, USA, ISBN 0471279110.

Deming W.E. (1986) *Out of the Crisis*. MIT Center for Advanced Education Service, Massachusetts, USA, ISBN 0911379010.

Hammer M. and Champy J. (1993) *Reengineering the Corporation: a Manifesto for Business Revolution*. Brealey Publishing, London, UK, ISBN 1857880293.

Hill T. (2000) *Manufacturing Strategy Text and Cases*. 2nd Edition, The Macmillan Press, London, UK, ISBN 0333762223.

Imai M. (1986) *Kaizen: the Key to Japan's Competitive Success*. Random House Business Division, New York, USA, ISBN 0394551869.

Parker M. (1990) Managing Successful Implementations (in *Managing Information Systems for Profit*. edited by Lincoln T.). John Wiley and Sons, New York, USA, ISBN 0471925543, pp. 147–197.

Index

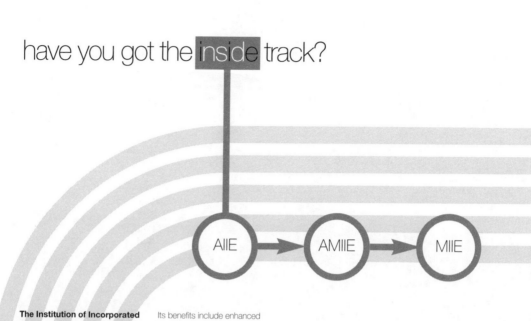